Dragon History

The Secret History of an Ancient Bloodline

(The History of Dragon Legends and Folk Tales Around the World)

James Rush

Published By **Hailey Leigh**

James Rush

Dragon History: The Secret History of an Ancient Bloodline (The History of Dragon Legends and Folk Tales Around the World)

ISBN 978-0-9952939-5-3

Legal & Disclaimer

The information contained in this book is not designed to replace or take the place of any form of medicine or professional medical advice. The information in this book has been provided for educational & entertainment purposes only.

The information contained in this book has been compiled from sources deemed reliable, and it is accurate to the best of the Author's knowledge; however, the Author cannot guarantee its accuracy and validity and cannot be held liable for any errors or omissions. Changes are periodically made to this book. You must consult your doctor or get professional medical advice before using any of the suggested remedies, techniques, or information in this book.

Table Of Contents

Chapter 1: The Monkey King

Sun Wukong is referred to as the Monkey King in Chinese legend. Even even though the traditional Chinese ebook Journey to the West made the monkey god famous in the 1600s, recollections about him have been advocated in Southeast Asia over the past three, 000 years. Sun Wukong has turn out to be appeared to be a monkey who finished suggestions. He brought on a wonderful deal problem inside the heavenly worlds but determined enlightenment via supporting others. In Chinese legend, Sun Wukong constitutes a effective enemy. He had lots energy that he may also moreover need to run as speedy as a falling well-known man or woman and hold up mountains on the identical time. He also can somersault 54,000 km consistent with hour. He can also freeze human beings and creatures in region and exchange the weather, that has been each very effective skill.

Sun Wukong's most frightening capacity have turn out to be that he must trade into any of the seventy- Earthly bureaucracy at will. This permits him trade into severa creatures and topics. Sun Wukong grows to be an professional in lots of preventing methods and will beat the super warriors in the celestial forces. Even his fur became magical. He may additionally additionally need to make copies of himself to assist him fight, or he can also utilize his fur to trade into weapons and special animals. Historians suppose Sun Wukong grow to be based totally totally on a aggregate of myths from antique Southeast Asian cultures, specifically the ones of China and India. The Hindu monkey discern Hanuman, written approximately inside the Sanskrit Ramayana, end up the number one element that long-established the person of the Monkey King.

As Indian Buddhists moved to China, they introduced their stories, which blended with Chinese customs and changed through the years In Journey Toward the West, Sun Wukong have grow to be as soon as a stone that got here to life whilst a breeze blew earlier than turning into the king of the monkeys. The god of wind changed into the father of the Indian ape god Hanuman. The Monkey King became furthermore stimulated with the aid of stories approximately the Chu Kingdom in 700 BC. Some testimonies approximately a monkey god got here from the Han Dynasty, and all of them ultimately stimulated Sun Wukong.

People love Sun Wukong and see him as a cultural hero, but Buddhist clergymen do not see him as a non secular determine. Instead, they see the Monkey King as more of a literary and creative character. Since the monarch of monkeys changed right into a rock earlier than he have come to be someone, Sun Wukong lacks any cousins or children. Because of this, Chinese religions do no longer honour him as an ancestor. Before Sun Wukong reached the Jade Emperor's high courts, the Monkey King have emerge as in charge of a fixed of apes dwelling in the jungles.

The Monkey King represents the definitely Chinese god who looks as if a monkey, so Sun Wukong represents one of the easy Chinese characters to tell apart. Before he have become enlightened, Sun Wukong frequently seemed as a bare monkey in photos of the Monkey King. After being let loose of his mountain jail and becoming a pupil of the Buddhist collect Tang Sanzang, Wukong Sun is set up wearing their Monkey King's crown, a gold chainmail blouse, cloud taking walks shoes, and the Monkey King's workforce, which weighs eight heaps however can be contracted right proper right down to the scale of a pin needle.

The symbols for vicinity, "awake," and "grandson" make up the call Sun Wukong.

But in this situation, the person solar way monkey. So, the real which means that of the moniker Sun Wukong is "a monkey woken up from nothingness." This ties in with one of the essential ideas of Journey to the West, this is that humans circulate from lack of awareness to records. People bypass from a mean-active prankster monkey who motives terrible issues to a smart Buddha.

In Chinese lore, Sun Wukong is referred to as the Monkey King. The tale of the monkey king have become first recommended inside the Song Dynasty ebook Journey to the West. In the e-book, the deceitful Monkey King, who thinks he's as properly as the gods for heaven, becomes the bodyguard of a Buddhist monk after being freed from a 500-yr prison below a holy mountain for stealing and eating the Peaches for Immortality and leading an riot in competition to heaven.

In the ebook Journey closer to the West, a sturdy magic stone modified into on pinnacle of the Mountain for Flower and Fruit. The mineral had a paranormal belly that could absorb yang strength from the heavens and yin power from the earth. One day, the magic rock gave transport to a small stone egg. And while the wind touched the magic rock egg, it come to be a stone ape that could go with the flow and communicate. When the stone ape opened its eyes, golden illumination shot into the sky and the Jade Emperor's fort. When the Jade Emperor noticed the mild, he despatched human beings to locate in which it got here from. When they advised him it end up just a stone monkey, he stopped annoying and concept it come to be not anything unique.

As the stone monkey commenced out investigating his new surroundings, he joined a troop of wild monkeys who frequented a move and carried out there daily. One day, the primates determined to have a opposition to look who would possibly come to be the ruler of the monkeys if they'll discover in which the flow into originated. The stone monkey grow to be the number one to plunge into the river and swim down a waterfall, wherein he decided a cave and an iron bridge in advance than every person else. They fast set up their new home nearby after persuading the opposite primates to sign up for him. The stone monkey informs the alternative monkeys of their vow and exhibits that he has continuously aspired to be known as the Gorgeous Monkey King.

One of the Monkey King's closest partners handed away after he had ruled for some time. Sun Wukong turn out to be disappointed with the idea of life and the ache that accompanies it, so he were given proper down to locate an eternal entity that could educate him on how to conquer loss of lifestyles. Sun Wukong seemed throughout the entire island he lived on. However, he modified into not able to find out any immortals who ought to offer him with answers to his questions. The Monkey King determined to wasteland his united states of america, so he built a raft and set sail at some degree within the water.

When he arrived on land, the individuals who discovered him concept he was some monkey-humanoid monster, in order that they fled from him as brief as feasible. Upon figuring out that he may also require a conceal, he grabbed some of the garb hanging outdoor to dry. He started out searching out immortality in the surrounding towns and villages, ensuring to cover his face. As he persevered his adventure, he have turn out to be privy to the corrupt nature of humankind. He decided to go into the woods inside the preference of locating an immortal. After taking walks, he placed out an immortal being resided at a neighbouring temple. After making a few inquiries on the temple, the resident martial artist, Puti Zhushi, denied him access after he asked permission to enter. Sun Wukong remained undeterred, so he sat on the temple's steps and pledged not to transport away except conventional to go into.

The Monkey King remained immobile on the steps for a excellent time. Because Puti Zhushi have turn out to be so focused on Sun Wukong, he invited the Monkey King within the sacred space and started schooling him as a disciple. It have become out that the Monkey King have become a geared up student, as he must draw close profound Taoist practices, alongside side the Path of Immortality—an extended length elapsed. Because of the incredible speed and skills of the king of monkeys, Puti Zhushi recommended Sun Wukong now not to show his abilties in the front of others, as doing so may want to make those humans desire Sun Wukong to train them. Students could get into problem if taught through the Monkey King, acknowledged for being a strict disciplinarian. People can also maintain that choice towards him if he decided on not to teach. Puti Zhushi then coerced the Monkey King into swearing that he may also want to never display the identification of his instructor.

The Monkey King woke within the woods, having formerly vowed to hold Puti Zhushi a thriller from all exclusive people. When he regarded round, he decided that no time had elapsed thinking about that he first stepped into the woodland, indicating that every one the training he had received turn out to be simply in a dream. When all people wondered Sun Wukong approximately wherein he had gotten his tremendous competencies, he would possibly say that he had decided out it all in a dream. He would in no way display the identification of his Master. Sun Wukong concluded that he wanted an enchanted weapon, given his wonderful capabilities. In the direction of his search for a mystical weapon, he decided himself inside the undersea fortress of Ao Guang, a few other of the Dragon Kings. At the palace front, Sun Wukong requested to be proven to the dragon's King, and the guards did their brilliant to prevent him from coming into. The Monkey King refused to offer in to their wishes and insisted that the ruler of the dragon want to in no way refuse the request of a brother king. He stated this to justify his function.

Following his stumble upon with Ao Guang, Sun Wukong requested a weapon appropriate for his abilties. As quickly due to the fact the historical Dragon King found out that the king of monkeys were as an alternative effective, he commanded that a massive kind of weaponry be introduced to Sun Wukong to attempt. The Monkey King in the end determined to use a body of workers referred to as the Ruyi Jingu Bang or Ding Hai Shen Zhen, which interprets as "the balancer of the four seas."It emerge as one in every of Ao Guang's most prized possessions due to the reality no individual else but the Monkey King possessed the essential power to wield the sword. The Dragon King is so impressed through manner of Sun Wukong's expertise with the personnel that he offers it to him as a gift. After this, the personnel earned the nickname "the group of workers belonging to the monkey king" due to its association with the King. The group of workers weighed almost 9 thousands and will independently trade dimensions, fly, and combat foes without its Master's assist. In addition, it have become viable to compress it to healthful in the area of a needle while it changed into no longer getting used.

The Dragon monarch then issued a summons to his other siblings, teaching them to procure the Monkey King royal clothing appropriate for a monarch. One most of the jewels have become a hat built using the wings of a phoenix, and it have become to end up famous as the "Crown of the Monkey King." In addition to that, he became furnished with a blouse embellished with gold chain mail and boots manufactured from cloudwalking cloth. However, the Dragon Kings had been no longer glad that objects have been given to the Monkey King given that they believed they had been being blackmailed in this manner. Following his departure from the kingdom set up through the Dragon King, the Monkey King made his manner another time to his mountain stronghold. Others have turn out to be aware about the Monkey King's electricity after demonstrating his newly acquired skills to his tribe. Soon after that, the Monkey King joined forces alongside the Saurian Demon King, the Bull Demon the King, a Single-Horned The Demon King, a Roc Demon King, the Lion the Spirit King, a Macaque Spirit King, as well as the Snub-Nosed Monkey the Spirit King.

However, unknown to the Monkey King, the Dragon Kings were asking heaven for retribution after their defeat on the palms of the Monkey King. Two messengers visited the Monkey King while he have come to be sound asleep to take his soul to Hell. Despite this, the Monkey King went to King Yama. He had his name, on the aspect of each exceptional monkey he have come to be acquainted with, removed from the Book of Life and Death. Infuriated, King Yama with the Dragon Kings rushed to the Jade Emperor, but the Jade Emperor's heavenly troops had been no in form for the prowess of Sun Wukong. The Jade Emperor welcomed Sun Wukong into his palace. He bestowed upon him the heavenly call of Protector for the Horses. He did this because of the reality he believed that with the aid of bestowing this form of end up aware of upon him, the king of the monkeys will be delivered beneath manage. Nevertheless, whilst the king of monkeys placed that Protector for the Horses

14

modified into his lowest rank within the celestial realm, Sun Wukong freed the horses, lower decrease back to Earth, and cited himself as Great Sage, Heaven's Equal.

Infuriated, the Emperor of Jade had clearly all started out to hassle an order to release an assault at the same time as he turn out to be endorsed to hold with severe warning due to the fact it'd be unexpected if the tyrant of monkeys were defeated. But things should skip incorrect if Sun Wukong have been to overcome the forces of heaven. Instead, it became cautioned to the Jade Emperor that he widely recognized the self-proclaimed name of the Monkey King but that he deliver them over again to the imperial house so that he may also need to cause fewer problems on Earth. The perceive of the Monkey King became definitely pointless. Within the heavenly realm, it emerge as seen to be a comic story regarding the Monkey King's lack of expertise regarding the significance of the region of heaven.

When an envoy arrived and advised the ruler of monkeys that Sun Wukong end up being extended to the Guardian of Peaches of Immortality name, Sun Wukong speedy normal the place. The change pleased Sun Wukong, and even as the envoy knowledgeable this Monkey King that Sun Wukong emerge as being multiplied to the area, Sun Wukong turn out to be thrilled. Nevertheless, as fast as he approached a peach grove, the king of monkeys couldn't face up to the appeal of ingesting off the peach trees. Soon after, he located that the monarchy in the West had dispatched her maidens to accumulate peaches to serve at the Peach Banquet. This is the event at which the Queen Mother bestows the present of persisted immortality on all celestial beings. Sun Wukong shriveled and wriggled interior a peach to break out them, discovering that he had actually consumed the excellent peaches.

When hiding inside the peach, he overheard girls giggling approximately how the king of monkeys become truly an immortal who gazed over the peach garden and wasn't a god. This became the on the spot even as he eventually discovered out the truth. Even worse, he wasn't even at the visitor list for the Peach Banquet! The Monkey King fled the grove in a rage and crept into the night meal hall, wherein he helped himself with the foods and drinks. Then, due to the fact the celestial beings made their way to the dinner, Sun Wukong travelled round heaven's distinctive tiers and halls. After Sun Wukong located that he had reached the level of the Dou Shuai palace at the top of the thirty third stratum, he stole their Pills of Being Immortal, took more in their Peaches of Immortality, took the rest of the king's wine, and then ran away to his mountainous u . S . A . On Earth.

This time, the Jade Emperor declared a whole-scale warfare in opposition to the Monkey King. However, Sun Wukong modified into now as an opportunity effective and had recently destroyed the 100,000 powerful celestial stress that blanketed the Jade Emperor, all 28 stars of the sky, and the four heavenly monarchs. Sun Wukong had even battled Nezha and Erlang Shen with a draw. Only via the efforts of Taoists and the Buddha, in addition to the goddess Guanyin, the ruler of monkeys have become eventually apprehended and condemned to be confined in Laozi's Eight-Way Trigram for forty nine days. This becomes the punishment for the Monkey King. Despite this, the Monkey King managed to break loose on day forty seven, the time he spent imprisoned, bestowing him a contemporary ability called the golden look. After he completed destroying his Trigram, he confronted the Jade Emperor.

The Jade Emperor awaited Sun Wukong to go back once more face-to-face with him in advance than straight away attractive to the Buddha. As the king of monkeys made actions in the palace, he requested that he be named the Jade Emperor because of the truth the forces of heaven couldn't defeat him, and he couldn't be held captive with the aid of any prison. Buddha, as an alternative, had a approach. He proposed a guess due to the fact the Buddha knew his exceptional flaw changed into the Monkey King's overconfidence. The Buddha placed a bet that the king of monkeys won't be capable of unfastened himself from the Buddha's palm. When he decided out he need to win, the Monkey King leapt from the clouds and soared to the arena's prevent. After counting only five columns, he decided to mark them the usage of his urine. Then he proclaimed himself a Great Sage on par with Heaven.

When Sun Wukong eventually again to the Buddha, he modified into prepared to ascend to the throne. On the possibility component, the Buddha disclosed that the five foundations he had exceptional have been actually the hands on his hand. Before the chief of the monkeys can also additionally want to interrupt out, the Buddha twisted his hand and despatched him hurtling in the direction of the earth, in which he come to be entombed beneath a mountain that the Buddha had sealed through a paper talisman and could not america for some other 500 years. This have become performed surely so Monkey King may also have a look at staying electricity and humility.

Chapter 2: Chinese Phoenix

Feng-Huang comes from legendary birds that had been essential in old Chinese cosmology. The Feng-Huang, or the Chinese Phoenix, signifies summer time and highbrow stability. It is one of the maximum revered animals in Chinese way of life, alongside side the dragon, the qilin, and the tortoise. People say that when it suggests up, it technique a large event will take place or that a boss is extraordinary. The Feng-Huang is regularly seen because of the truth the religious union of male and female, with Feng being male and Huang being girl. It also can be seen as a girl spirit while linked to the male dragon. So, images regarding Feng-Huang and the dragon together can recommend both combating or peace within the marriage.

The Feng-Huang is a symbol of all of the fantastic things in lifestyles. It is lovely, fashionable, dependable, and sincere. It calls us to a better level of existence, entire of justice, peace, boom, and religion. Something like this doesn't need a herbal counterpart; the Feng-Huang possesses the satisfactory of the natural global in its snap shots. People are people who want to make the ones ideals actual on the earth. Feng-Huang incorporates one male hen and one female fowl, even though it looks like a unmarried being. Both ladies and men are tested, however Feng is the male, and Huang is the lady.

People say that the Feng-Huang includes a crow's beak, a swallow's face, a hen's brow, a snake's neck, a goose's breast, a tortoise's once more, a stag's hindquarters, and a fish's tail. Its body stands for the six heavenly our our bodies. Its eyes represent the solar, its decrease decrease again represents the moon, its wings are the breeze, its feet are the ground, and its tail is all the first-rate planets.

The Phoenix inside the Western World dies and returns to lifestyles, but the Feng-

Huang does not need to. It lives all the time. The chook likes music and builds its nest high inside the K'unlun Range range. This holy creature might now not kill. It high-quality eats flora and in no way bugs which might be although alive. It first-class eats bamboo seed and lives in paulownia wood, its natural home. It in no way hurts flowers for no reason.

It's not clean where the Feng-Huang story got here from. Some college students anticipate it is probably a photo of a big chook from the past, like an ostrich, that have become not unusual in historic China. This is much like the concept that dragons had been legendary creatures that regarded like dinosaurs. The Feng-Huang first seemed in Chinese society around 3000 BCE, genuinely in advance than the Yellow Emperor died. The legend says that the Feng-Huang high-quality shows up at the begin of a modern-day age and hides at the equal time as subjects go wrong. It can show up for the duration of instances of peace and plenty or even as a amazing chief is born. The hen comes down from heaven for the earth due to the fact the signal of a new technology and the begin of plenty. The bird represented a time of peace and wealth when a modern emperor took energy. It have become moreover often used to symbolize the Empress.

Feng-Huang have grow to be an vital part of China's vintage cosmology, which said that the world and heavens have been made through the dragon, the qilin, the tortoise, and Feng-Huang. The international turned into split into 4 components, and the Feng-Huang have become in fee of the Southern Heaven vicinity, which stands for summer time. The Feng-Huang and the roaring beast are sometimes proven collectively as enemies and every so often as glad fanatics.

Over the years, the Feng-Huang have become seen as a signal of power, wealth, beauty, and goodness. People suppose that

Feng-Huang's appearance is a exceptional signal that a brand new age in records has all started out. People say, "They excellent show up in places wherein there can be peace, wealth, and happiness; even as matters pass incorrect, they run to the heavens."Anyone who sees it unfold its wings will understand that it's miles right. It stands for equity and humanity, teaches all of us a manner to do the right issue, and inspires religion in anybody who sees it. The Feng-Huang changed into sometimes visible due to the fact the holy sign of the union of male and girl, that is what maximum humans name the Yin-Yang picture. The Yin changed into female, and the Yang modified into male. When they arrive collectively, like while Feng and Huang come together, they display how contrary forces can create stability and concord.

In the royal palaces, the Feng-Huang end up visible as a photograph of the holy woman and the Empress. On the opportunity hand, the dragon stood for the Emperor, and paintings regularly showed the dragon chasing a Feng-Huang or the 2 animals together. The maximum common photograph nowadays is this one. However, it is crucial to endure in thoughts that Feng-Huang stands for the union of a person and a lady. People concept the Feng-Huang managed the five tones of Chinese track and stood for the Confucian ideals of justice, honesty, loyalty, and decorum. The Feng-Huang tested that the those who lived in a house were honest and committed when they had been used to designing it.

For hundreds of years, the Feng-huang turned into hired in Chinese paintings. When it grow to be used as a house decoration, it normally supposed that the folks who lived there were honest and might be right to their buddies and family. It modified into often made from Jade and worn as rings or a sign of top achievement. People who wore Feng-huang earrings had been taken into consideration very upright, so excellent a small group were allowed to non-public such earrings. In the surrender, the bird changed into moreover decided on coffins and graves to expose that the person have been precise in lifestyles. Many art work and work of artwork of the feng-huang showed it preventing the snake, its inherent enemy, or with the 3 greater divine creatures from vintage Chinese mythology.

Yeren: Chinese Bigfoot

The Yeren is a Bigfoot-like creature this is local to China. It is likewise noted by using the names Yiren, Yeh Ren, Chinese Wildman, and Man-Monkey. The Yeren is a exquisite primate this is concept to are residing inside the highlands of China, with maximum noted sightings originating from the remoted area of Hubei. Although sightings of white Yeren are unusual, they had been said; this can mean albinism in older members of the family or a massive jaw in extra younger people. The Yeren's hair has a reddish brown colour. It can achieve heights of up to 8 feet, and its way spherical people levels from exceptional to reserved. It has been stated that extra than four hundred human beings have seen the Yeren, as said via Xinhua.

The Chinese authorities have achieved enormous searches all through the united states of a to discover the Yeren. There have been a superb form of footprints and hair traces placed. However, much like the Sasquatch or the Yeti, the life of this species has in no way been tested through way of technological knowledge. Therefore, it maintains to exist within the realm of fantasy and cryptozoology. Despite its stature, a few bills describe it as a good deal plenty much less long lasting and stocky than its cherished ones, which includes the Sasquatch. According to the recollections of the intended monster, the yearn walks erect and is over 2 meters tall; it's miles covered in tawny hair across the frame, fairly lengthy on the scalp, and it has a face this is suggestive of every an ape and a human. Other ordinary tendencies include hair that is black-red, eyes which can be swollen, extended hands that hold close proper all the manner all the way down to the knees, and big toes. There is a common notion that the yeren will snort while encountering someone.

In nearly all cases, the eyewitnesses do now not describe the animals as having any reddish-colored hair on them. There have additionally been opinions of seeing a few white specimens. Their height is predicted to variety among six and 8 ft, even as some big examples purportedly recognition over twelve feet tall have been said. Their weight is projected to vary from to 4 pounds. The Yeren has a depressed face, protruding lips hiding first rate, horse-like teeth, and a large nose with upturned nostrils, all of which make contributions to its head form, that's extra similar to humans than one-of-a-kind apes. It is commonly drastically a splendid deal much less huge than the Bigfoot visible in the United States.

The Yeren, just like Bigfoot, is a slight creature that, at the same time as it comes into contact with humans inside the Zhejiang province, will generally walk away silently regardless of the large notion that Yeren is a peaceful being. A report posted in 1980 unique the ordeal of a girl who stated that she have been kidnapped via this type of beings. She stayed in the Hubei province for twenty-seven days on the identical time as the extraordinary chimpanzee managed to get her pregnant. The offspring handed away at 22, a decade in advance than the document modified into published, and a next studies of his bones allegedly showed tendencies of every guy and ape!

Researchers that have a observe cryptozoology have hooked up a connection the numerous Yeren and the possibly extinct pongid Gigantopithecus, which used to stay inside the identical desired place. Another idea proposes that the Yeren is a mistake of the gibbon ape, that's severely endangered. It has additionally been proposed that the Yeren is, in reality, a wholly novel sort of orangutan, one that lives at the ground, walks on legs, and is indigenous to mainland Asia in place of Borneo or Sumatra. There is also the possibility that the Yeren and one of a type similar cryptids similar to the Yeti and the Sasquatch are in detail associated with people.

Scientific interest in such apemen exploded within the 1950s and Nineteen Sixties with pseudoscientific unearths concerning Bigfoot and the yeti. However, pressure from the Maoist government to move away those myths and folks memories inside the beyond repressed extra hobby inside the Yeren until its dissolution in 1976. After that, the Chinese Academy of Sciences organized massive expeditions to analyze meant eyewitness recollections, footprints, hairs, or our our bodies as "Yeren fever" took preserve with out scientists working with an tremendous dependence on citizen studies. These expeditions were initiated at a time at the same time as " Yeren fever" had taken hold. It have become typically believed that the Yeren grow to be a human ancestor that lived extended inside the beyond, along with Gigantopithecus and Paranthropus robustus. The notable evidence of the monster came from famous creatures like bears, monkeys, and gibbons. By the past due 1980s, the medical community had misplaced hobby within the concern. Despite this, there's however ongoing organized research on Yeren, however the reality that no credible scientific establishments renowned such apemen.

In Chinese folklore, memories approximately "wild men" and amazing animals with comparable developments were surpassed down orally and in written form for millennia. They might also have made their first look in written shape inside the Jiu Ge, written thru Qu Yuan, who lived in Chu within the Warring States duration. Qu Yuan flourished from 340 to 278 BCE. The time period "mountain spirit" appears in His Ninth Song; in Chinese literature, such characters nearly regularly allude to a man or woman. Depending on whom you ask, the Yaoguai, the ogre, or a humanoid clothed in a fig leaf can also need to constitute the mountain spirit. Other interpretations consist of an ogre. In 1982, a Chinese paleoanthropologist named Zhou Guoxing unearthed a lantern that become 2,000 years antique and had an adornment that appeared to portray a "furry guy." This discovery moreover relates to an ancient tale of untamed guys. By the past due Nineteen Eighties, medical hobby had all. However, it evaporated due to the fact not one of the missions have been a fulfillment in unearthing convincing proof. The stated bodies, hairs, and footprints originated from hundreds of extraordinary recognized creatures. These animals covered people, Himalayan bears, Tibetan bears, macaques, gorals, and serows. The so-referred to as "monkey boy" skulls, which have been notion to be proof of Yeren and human hybrids, had been shown to have belonged to truly human children who suffered from spinocerebellar ataxia. Eyewitnesses might also moreover moreover have erroneously recognized bears, gibbons, or monkeys because of the fact those sightings are typically encouraged at a distance. In addition, it is quite feasible that severa eyewitness memories had been thoroughly made up or exaggerated.

Chapter 3: Tiger With More Than One Wings

Qiongqi is one of the Four Fiends referenced in ancient Chinese mythology. This mythology is in particular recorded within the conventional paintings "Shan Hai Jing," additionally referred to as "The Classic in Mountains and Seas." A description of Qiongqi may be discovered in the part of the text titled "Inner Sea Classic." It depicts Qiongqi as having a tiger with a fixed of wings, and it's far well-known for its propensity to eat people, beginning with their heads. This beast is a fierce and scary one in its very very very own proper. It's thrilling to be aware that the precise text consists of a second tale of Qiongqi, which may be determined in the "Western Mountains Classic" segment.

In this tale, Qiongqi is depicted as a bull with spikey hair, that is a terrific departure from the preceding instance of the creature within the "Inner Sea Classic." However, each variations of the creature percent the common function of being hungry guy-ingesting creatures. Regarding this issue, there is no distinction among the 2 of them.

The implementing nature of the Qiongqi as a monster is emphasized in numerous legends surpassed down through Chinese life-style. The reality that it's miles covered inside the "Shan Hai Jing" highlights the troubles and dangers that historical Chinese society grow to be up towards. The ever-present danger that Qiongqi poses serves as a regular reminder to human beings that it is vital to keep vigilance and shield moral necessities. Although unique money owed may also describe the arrival of the Qiongqi in a different way, the essential nature of this terrifying creature has not altered. Its appearance in Chinese mythology demonstrates the big and charming universe of mythical beings, which have served to amuse people and train them to be careful of their surroundings through the years.

In modern society, Qiongqi may be referenced and portrayed in diverse cultural genres, consisting of literature, artwork, and new retellings of traditional reminiscences. Its function in Chinese mythology brings to light the numerous and fascinating vicinity of mythological creatures, that have served to entertain and educate vital life commands at some stage in information.

In the "Inner Sea Classic" a part of the "Classic of Mountains and Seas," Qiongqi is described as having the size of a cow and the appearance of a tiger. It has a couple of monster wings and a starving appetite for flesh from people, beginning its meal with the victim's head. It kills its patients thru biting their heads off. Because of this, the creature is each terrifying and ugly in look. In contrast, the "Inner Sea Classic" depicts Qiongqi as having the arrival of a bull with spikey hair, demonstrating big variations from the sooner story.

Other writings describe Qiongqi as having snow-white fur, moderate-gold dragon horns on his forehead, antlers reminiscent of a bald eagle, and an array of black wings. In addition, Qiongqi is depicted as having black wings. Although the veracity of those bills is questionable, it seems that they percent the feature of having a desire for consuming human flesh. Despite this, the person of Qiongqi as a monster that devours humans is emphasised inside the three interpretations, with moderate variations among them.

Traditional Chinese lifestyle imbues the mythical beast Qiongqi with profound connotations because of its reputation as a cultural icon. It is concept that the Qiongqi can absorb nutrients and research into its "stomach," it certainly is symbolic of its functionality to digest human information and know-how inner itself. This notion is primarily based mostly on the Qiongqi's considered certainly one of a kind trait of getting a mouth under its stomach. This is regular with its portrayal as a species that could talk with human beings using human language.

Qiongqi is infamous for its blatant contempt for righteous ideas and has a propensity to take the element of the dishonest in region of the righteous. It is thought to step in whilst disagreements or conflicts rise up. Yet, the results of its interventions invariably counter the moral requirements that individuals uphold. Qiongqi has a propensity to assist the factor that is unjust whilst persecuting the facet that is certainly. For example, it's far viable to devour the nose of a person who is not responsible or even to deliver untamed animals to a criminal who has perpetrated lousy crimes.

In addition to this, a few views see Qiongqi as a metaphor for intellectual notion. By consuming each residing detail over an uncountable length, Qiongqi exemplifies the concept of the "Great Dao" or the countless boundary. It represents an in no way-finishing quest for the last truth that underpins the entirety.

According to Chinese folklore, the Qiongqi is the most nimble of the four terrifying monsters. It has dragon wings on its lower back and the ft like two dragons. It is concept that after the passing of Gonggong, it underwent a change and now holds the potential to exert control over aquatic species. Even although it's miles classified as an animal, it's miles very just like the dragon species. Qiongqi can assume leadership over the world's dragons and update the Four Sea Monster Kings with the aid of the use of bringing clouds and rain into existence.

In the historical Chinese feng shui method, the Qiongqi is a mythical beast this is stated to play an crucial function in Chinese geomancy. It is sometimes portrayed as a fearsome and robust creature with the frame of a tiger and wings like the ones of a hen. In feng shui, "Qiongqi" refers to offering protection and fighting off terrible power. It is concept that Qiongqi possesses a effective and watchful electricity that can protect closer to risky energies and evil spirits. This strength functions as a protector. It is regularly hired as an image of symbolism or talisman to defend homes, groups, and people in the direction of unfavorable forces, ill fortune, misfortune, and exceptional styles of calamity.

In feng shui programs, Qiongqi sculptures or collectible collectible collectible figurines can be positioned strategically near doors, domestic home windows, or important portions of an area to function as a protecting presence. It is said that they might instil a feel of protection, protect in the direction of the go with the flow of poor chi (power), and make the environment more harmonious and super.

It is connected to engaging in achievement regardless of difficulties, rising to satisfy worrying situations, and maintaining up a stalwart defence in opposition to adversaries. People cause to nurture inner strength, boom their self guarantee, and manual their functionality to conquer challenges in every their non-public and expert lives via manner of using evoking the strength of Qiongqi. This is completed in the hopes that it will assist them. It is critical to keep in thoughts that the best positioning and use of Qiongqi within the principles of feng shui practices might range from one practitioner to the subsequent, depending at the specific targets or intents of the feng shui layout. In the equal way that it's miles encouraged to are searching for for recommendation from an professional feng shui practitioner earlier than the use of any feng shui remedy or picture, it's far outstanding to perform that even as searching for specific assistance and advice.

"The Records via the Grand Historian: Annals of the Five Emperors" documents the beginnings of Qiongqi thru noting, "There changed into an incompetent character from the Shao Hao extended own family who slandered special people, destroyed bear in mind, and unfold evil phrases." The humans of the region gave him the decision Qiongqi. He, together with "Hundun" from the Hong extended family, "Taowu" from the Zhuanxu prolonged own family, and "Taotie" from the Jinyun clan, are together known as the "Four Demons." Emperor Shun exiled them and "sent them to the four frontiers to govern evil spirits."

Huang E modified into Shao Hao's mother, and her father become traditionally known as the Yellow Emperor. Other names for Shao Hao include the Western Heavenly Emperor and the White Emperor. The "Records of Lost Chronicles" tells their story, that is lovely and shifting and may be found in that ebook. According to the legend, the

Qiong mulberry tree exists, wherein the fruit matures simplest once each 10,000 years and ingesting it bestows the existing of limitless young adults at the client. These adjectives, which include "white" and "gold", endorse that Shao Hao's ancestry dwelt in the Western vicinity. Shao Hao, furthermore referred to as Shao Hao, presided over the vicinity of the West, moreover referred to as the "Qiong Mulberry extended family" or the "Golden Heaven prolonged circle of relatives," however greater normally called the "Qingyang prolonged circle of relatives." After being chased out thru the Shun, Qiongqi have end up pushed to the northwestern region.

In a long term lengthy ago, the water deity called Gonggong modified into notorious for his haughty behaviour, boasting to all of us that he modified into 2nd in power best to the heavens however unconquerable within the international. After being attentive to this, Zhuanxu, the ruler of the Sacred Flame, scoffed at Gonggong's stupidity and dispatched his minion Zhu Rong to undertaking the water deity. Consequently, Gonggong and Zhu Rong were compelled to interact in bloody fight on Mount Buzhou that lasted seven days and nights, with neither growing because of the fact the unequivocal victor.

As someone certainly prideful, Gonggong was outraged via using the truth that Zhu Rong might also want to defeat him regardless of believing he turned into unstoppable. He lost his temper and slammed his brow in competition to Mount Buzhou, inflicting excessive damage. This collision brought catastrophe to the area of the mortals, which in flip delivered approximately the human beings of this worldwide to experience problems sometimes. As a quit end result of Mount Buzhou's crumble and the celestial river's overflow, huge destruction passed off, and innocent humans were harmed.

To our right fortune, Yu the Great arrived later to prevent the flooding. Yu, who led an army of humans, correctly eliminated the water-primarily based deity Gonggong. Despite this, Gonggong did no longer disappear even after he passed away. A vengeful and evil spirit from internal his frame erupted and converted into Qiongqi. Its roar looked like a dog barking, taking the shape of Qiongqi.

Chapter 4: Mythical Woman

China has a tale about a lady named Jingwei, who have become Yandi's daughter. She died in the Eastern Sea while she become more youthful. After she died, she emerge as a chook and flooded the ocean with sticks and stones to get returned at her killer. At one aspect, Jingwei tells the sea that she cannot fill it up in a million years. Then she says it will take her about ten million years to use it, and he or she'll do some aspect it takes to make sure nobody else dies. When she died, the girl's soul modified right into a chook with a flower head, a white mouth shell, and pink talons on its feet.

The ruler, Yan, is said to have a daughter called Nuwa. He cares plenty about the growing sun over the East China Sea and the manner it may heal. He goes there each morning to locate the solar. The female regularly played at the aspect of her dad at the equal time as he wasn't domestic. He have emerge as to take her out of doors to peer the dawn. She couldn't go along with her dad due to the truth he have come to be busy with exceptional paintings. The lady went to the East China Sea to take a look at cutting-edge dawn. The awful information is that the hurricane inside the location became the boat over, and the waves swept her away. Emperor Yan knew she have become sick, however there was no remedy he ought to use to maintain her. She is set on which include water to the East China Sea to make sure no man or woman else dies like she did. She takes rocks or sticks in her mouth each day and throws them into the water to fill it up. The Eastern Sea says she can't do it, and it's going to take her one million years to fill it up. Jingwei, who become very cussed, spoke back that she ought to fill the sea with water for ten million years simply so nobody else would be pressured to perish as she did.

Sanzuwu: Three Legged Raven

In Chinese lifestyle and mythology, the 3-legged raven is a sign. It's furthermore mentioned in masses of reminiscences and regularly proven in vintage Chinese artwork. The sanzuwu disc stands for the sun. From historic instances, this appeal changed into used to make garments for the emperor. A sunbird named the Yangwu and the Jinwu is the most commonplace manner to attract a Sanzuwa. Even even though it's referred to as a crow or raven, it's far commonly pink in place of black. It's furthermore demonstrated as an illuminated crow in some digs. Another commonplace picture of the Queen Mother for the West has 3 dragons, a tom cat, and a manservant surrounding her. A leaping frog and a 3-legged fowl also are spherical her.

People used to count on that the three-legged raven changed into the handiest who added the Sun out of inside the back of the clouds. The ten sun crows had been stated to be accountable for drawing the sun out.

Their task modified into to fly off and make manner for the alternative crows. Two sorts of vegetation that would most effective be decided on this planet have been the crows' preferred food. Some of those crows should take off to eat them, changing the solar's flight path. Xihe, the mother of the institution of ten crows, did not like how her children stored converting the sun's route. They couldn't fly down and munch their grass because she made them blind. They all decided on to fly away subsequently and burn the area down. Another god, Houyi, who modified into a heavenly archer, killed every person however one maximum of the crows. There was handiest one crow left within the world with three legs.

Many assume a sanzuwu looks as if a sun crow named the Yangwu, moreover known as the Jīnwū, because of this that that "golden crow." It is generally pink in area of black, although it is referred to as a crow or raven. At a Mawangdui archaeological internet web site on-line, a Western Han silk picture moreover indicates a "golden crow" within the solar. According to historic Chinese artwork, Fuxi, the god of advent, is confirmed preserving the solar's disk with the jīnwū. Nüwa, the lady of advent, is validated keeping the moon disk with a gold-striped toad. Folklore says there had been ten solar crows in advance than everything, and that they settled in ten unique suns. These humans sat on a crimson mulberry tree in the east, at the bottom of the Valley of the Sun. The tree's name, Fusang, approach "the leaning mulberry tree." People stated that this mulberry tree's branches had many mouths that opened. Every day, Xihe, the "mother" of the suns, have to pick out out one of the sun's crows to adventure around the world on a carriage.

One solar crow may want to go back, and each different would go away on its adventure at some stage in the sky. The solar ravens loved eating more than one grasses of existence, which Shanhaijing known as the Diri (this means that "floor solar") and the Chunsheng (because of this "spring increase"). From time to time, suncrows need to come down from heaven and eat these vegetation. Xihe did no longer like this, so she placed her fingers over their eyes to prevent them. According to legend, round 2170 BC, the ten solar ravens got here out concurrently, setting the sector on fireplace. Houyi, the heavenly archer, spared the day through killing all however one of the solar crows.

The crow with 3 legs turn out to be called "Sanzuwu" in Chinese society. There were testimonies that this imaginary chicken changed into in price of the Sun's every day journey in some unspecified time in the future of the sky. Folklore says there as quickly as used to exist ten solar crows, and it become the technique of each crow to fly out and unfold the sun all through the sky.

Two varieties of grass that would most effective be determined on earth have been the crows' great food. Some of those crows would possibly take off downward to devour them, which modified the sun's flight direction.

That made Xihe, the "mom" of the 10 crows, indignant, so she blanketed all of their eyes to lead them to blind simply so they could no longer fly all the manner right down to devour their grass. One day, despite the fact that, all ten crows took off, which set the world on fireside. Some crows were killed thru some different god called Houyi, who've come to be a divine archer. There grow to be exceptional one crow left within the worldwide, with three legs. I find it exciting that crows in those myths aren't portrayed as evil, scary, or cunning like we usually recollect them. I positioned it interesting that a crow had been given this form of huge approach and became doing top matters for humans in place of being a terrible sign or caution.

Longma: Winged Horse With Dragon Scales

People have normally been inquisitive about many animals from mythology and folklore. One interesting example is the pony with wings and dragon scales; it has a horse's splendor and the dragon's power. Here, we will have a study this uncommon beast, its tremendous names, and the myths surrounding it. Animals that appear to be people are famous in mythology and folklore. The sphinx, the gryphon, and the centaur are only a few well-known examples of this. Another interesting creature that humans want to think about is the horse dragon, which is said to have the beauty of a horse and the power of a dragon.

The horse dragon has remarkable names in wonderful cultures and myths. As the call indicates, Longma method "dragon-horse." In Chinese lore, that is the name for the horse dragon. This extraordinary animal has a horse's body and head and a dragon's tail and wings. Many people concept the Longma had magical powers, like converting the climate and making it rain. Emperors and kings regularly rode them because of the fact humans noticed them as signs and symptoms and symptoms of authority and success.

What are a few other names for the horse dragon in Western mythology? Some of them are "dragon horse," "dragon steed," and "dragon mount." The Longma is a monster from Chinese mythology, however the dragon horse is a creature from Western folklore. An vintage trope is for knights or opponents to enjoy those frightening animals and rush into struggle. People have given the horse dragon many extraordinary meanings and images. The Longma is a beast for strength and actual achievement in Chinese folklore. Few human beings count on that seeing a Longma will beautify their life. On the alternative hand, the dragon horse in Western legend is often showed as a hero. The beast comes from myths about heroes and the values they stand for.

A legendary animal in China called the Longma is also once in a while described as a "dragon-horse." It is a fictional animal that resembles a horse and a dragon. It is just as frightening and fierce as either one. The Longma has the top and frame of a horse and wings and dragon scales in the course of its body. This creature is considered holy due to the truth it could fly, dive, and stroll on land. Because people concept it had supernatural powers, like controlling the weather and making it rain, it changed into often used as a signal of energy and splendid luck in paintings.

The Qin Dynasty emerge as the number one time that Longma have turn out to be made. During this time, the Longma have become appeared as a signal of wealth and success, and it was often proven as a powerful animal that rulers and kings rode. Many people notion it may trade the weather and the trails of rivers and mountain ranges. In Chinese legend, the Longma grow to be moreover essential. In The Dragon King's Daughter, a Longma falls in love with a princess and chooses to show right into a human so he may be in her. Longma's father, the Dragon King, does not like that his daughter is getting married, so he offers the princess many tests. A princess is subsequently prepared to marry the Longma after she gets beyond her troubles, and they live blissfully ever after.

People often mix up the Longma and the Kirin, creatures from Chinese mythology. With a horse's head, the Kirin looks like a move among an animal and a dragon. The kind creature is regularly demonstrated as a supply of happiness and real fulfillment. Heroic humans and emperors rode the

Longma into struggle as it modified right into a volatile animal.

People frequently assume that the Longma and the Kirin, testimonies about creatures from Chinese folklore, are the equal. These creatures are very specific from every specific except for the matters they have in not unusual. The Longma is a made-up animal with a horse's frame and head and a dragon's scales and wings. It's a frightening animal that kings and heroes often rode into battle. People perception it had magical powers, like regulating the climate and making it rain. Because of this, it changed into associated with energy, wealth, and well achievement.

Conversely, the Kirin has a deer's body, a dragon's scales, and a horse's head. The animal is content material and does now not cause any damage. People say it most effective shows up in the course of fantastic peace and wealth, so it's far frequently used as a sign of information and notion. The

Longma and the Kirin are very precise animals, even though they each have dragon scales. The Longma is a sign of strength and power, and the Kirin is an indication of precise fortune and records. The Longma's body is type of a horse's, and the Kirin's is type of a deer's.

In Chinese folklore, the Longma and Kirin are of the four heavenly beings. The special are called the phoenix and a turtle. Together, they constitute the look for balance and peace in the universe. The Longma and the Kirin are important animals in Chinese mythology, regardless of the reality that they serve particular competencies and constitute severa things. They preserve fascinating and motivating people everywhere inside the international.

The Longma is a horse with dragon scales and wings in Chinese folklore. It has been interior for quite some time. Because the Longma has such deep roots, it's miles tough to say how vintage it's miles. During the Qin Dynasty, the Longma have become first noted. Art showed the Longma as powerful animal rulers, and kings regularly rode. The Longma changed right into a signal of wealth and top suitable fortune in some unspecified time in the future of this time. Because it had supernatural powers, it have turn out to be an important image in Chinese lore, and those nevertheless use it these days. The Longma has been crucial in Chinese society and mythology for hundreds of years. Its reputation has grown out of doors of China manner to the numerous works of paintings, writing, and film that display it. Since the Middle Ages, the Longma has moved and stimulated people worldwide, together with those in China.

A unicorn and horse with wings and dragon scales are stated to be smooth and delightful. Even so, there may be a smooth difference a number of the two in terms of tactics they appearance and are raised. The Longma, moreover called the dragon-scaled flying horse, is a legendary horse in China that typically has wings and scales that appear to be dragons. People frequently connect it with power, appropriate success, and the supernatural, like being capable of change the go with the flow of rivers similarly to flow into mountains. Unicorns, conversely, are creatures from European legends that are thought to appear like horses with one horn on their forehead. People suppose it could smooth water and assist with many illnesses due to the truth it is connected to those specific features.

In artwork, every animals are confirmed to be glossy and beautiful. However, they're very top notch animals with unique cultural meanings. Compared with the unicorn, which stands for purity and healing, the Longma stands for electricity and supernatural capabilities. But the legendary unicorn and the flying horse in dragon scales are no matter the reality that validated in art, literature, and well-known manner of life as they fascinate and inspire people worldwide. In stop, humans have been curious about the mythical horse with wings and dragon scales for hundreds of years. In a few cultures or myths, it has a exceptional shape and this means that, but in maximum cases, it stands for energy, fulfillment, and top specific fortune. The Longma, additionally known as the dragon horse, is a famous animal that has prolonged stimulated humans international.

Chapter 5: One-Legged Mythical Bird

The Bi Fang is a legendary bird with exceptional one leg that looks in conventional Chinese mythology. Although there are numerous different techniques to represent it, it's miles most customarily portrayed as some form of crane. It is said that the fowl may be determined atop Mount Zhang'e, and its presence may be interpreted as an omen of an coming close to fireplace. However, it isn't a monster normally considered horrifying, and the Yellow Emperor as speedy as saved it as a accomplice. On the opposite hand, many claim that Bi Fang is the best who starts offevolved offevolved fires inside the adjoining organizations.

The Bifang is placed on China's mountainous and desolate Mount Zhang'e. It has the appearance of a crane however first-class has one leg; its beak is white, and it has red markings set toward a green background. The sound it makes is identical to its name.

A bifang changed into a signal that a city ought to quick be engulfed in flames for no apparent motive. This might be related to the shade pink that it has. However, it have come to be now not normally a portent of sick fortune, as it takes region as a useful attendant to the Yellow Thearch within the Master Hanfei and because the spiritual essence of timber within the epic Master of Huainan. Both of those appearances may be located within the Master in Huainan. According to severa payments, the bifang grow to be the only who commenced out the hearth via the use of the flames which have been stored in its mouth. Mathieu compares it to the Chinese crane, which has a dependancy of resting on one leg and can have inspired the manner the bifang seems.

Nian: Monster Dog With Two Long Horns

This is the Chinese phrase for "yr." China has a story about a beast referred to as Nian that lived within the hills or under the ocean. There are one-of-a-type tales approximately how the Nian monster seemed. Some say it resembles a dog with a flat face and distinguished the the the front teeth. Other human beings say the Nian monster has lengthy horns and quite a few sharp enamel, making it huge than an elephant. Its horn is supposed to attack its food.

People in China moreover assume the mythical Nian monster has frightening eyes. Nian is an unsettling creature that scares humans, however it's far frightened of the colour due to the fact it's so high-quality. People could probable located red paper decor on their the front doors and steamed buns at the home windows to scare Nian away on every occasion they knew he ought to attack on New Year's Day. People in China count on that Nian is fearful of fireside and pink. They spark off firecrackers and lighting all night to scare the monster away. Nian can't stand loud noises, both. That's why there are a whole lot of loud noises in a few unspecified time inside the destiny of the New Year activities.

An vintage time in the past in China, there was a dangerous animal known as Nian. The Ning monster came up from the sea's depths on the very last day of the lunar yr to go to the land of people who live. It arrived to consume pets, food, humans, and particularly children. Because humans knew of the nian beast, they could eat early on this day and near the gate to their animals. Then they hid in the woods some distance some distance from the giant Nian monster so it might no longer attack them.

One year, no matter the reality that, this modified at the same time as an elderly gentleman came to the town. His hair have become gray, and his pores and pores and pores and skin come to be crimson. A grandmother came as much as the elderly gentleman and provided him with food. She suggested them how scary the monster became and tried to get him to cowl in the back of her and the opportunity locals. On the alternative hand, the vintage man did now not seem to mind and saved his cool. He stated he need to get rid of the monster and requested to spend the night time time at his grandmother's residence.

The grandmother wasn't effective, however she did what he requested and ran to the mountain, leaving the vintage guy within the lower back of. Nyan, the monster, broke into the metropolis around midnight. He felt a few detail wonderful all of a shocking. In the past, whilst the monster showed up, the complete metropolis changed into dark. But this time, there has been moderate coming from an east-facing residence. The monster cautiously approached the house and noticed that crimson paper have been used to cover the home windows and doors. Also, the aged guy had lit many candles inside the residence. The beast gave those atypical subjects the evil eye and walked as a whole lot as the front door.

The antique guy inside the red dress came out while the door opened. The monster ran away because it grow to be scared. When they over again the following day, the city wasn't damaged. The vintage granny warned the opposite townspeople that the vintage guy have become determined to scare the animal. When all of us were given to the house, all the doorways and domestic home windows have been included in crimson paper, and lights lit internal. There modified into furthermore bamboo that hadn't been burned within the out of doors. After a few perception, they knew that the aged man became a god who got here to help them. He informed them approximately the hidden guns that would scare the Nian away. Things like firecrackers, flashing lights, and purple topics are on this organization.

The tale of the nian animal is an crucial a part of the Chinese New Year. The monster's name, nian, because of this three hundred and sixty five days, is the reason for this. New Year's Day is called Guo Nian, which means "celebrating the contemporary 12 months to get beyond Nian." There is also a connection among the Monster Nian and the Chinese New Year. People expect that the monster used to make its way to the village on the night time time in advance than the New Year to kill humans. So, the Chinese experience the vacation to bear in mind how they beat the monster.

The Nian dance is a few issue that human beings do every yr on the night time time in advance than the Chinese New Year. People although use crimson lanterns and loud firecrackers ultimately of the vacation due to the reality they anticipate those gadgets scare away the demonic Nian. In addition, they carry out lion dances at the same time as wearing mask and clothes that mirror the Nian. Now that humans apprehend the way to scare the monster, leaving the lights taking walks and last up past due to keep away from Nian has unfold to big elements of China. In Chinese life-style, this have become a huge tour now referred to as Lunar New Year's Eve.

You likely understand about the Nian Monster in case you ever celebrated Chinese New Year. This animal is risky and used to eat humans on New Year's Day. But after a three hundred and sixty five days, an elderly gentleman turn out to be able to frighten it away. The locals additionally discovered from him the manner to maintain the beast away. People have a laugh the New Year with loud noises, flames, and pink paper adorns because of the reality they assume they may scare away the demon.

Pixiu: Dragon Headed Lion Like Creature

In historic Chinese society, the Pixiu, Pi Yao, Tianlu, and Bixie had been large legendary creatures. People accept as true with it to be a lucky signal to supply them wealth, safety, and real good fortune. Because of this, it is often employed as a decoration, and you may see it in most homes and locations in China. The pixiu is usually a decide that stands shield. The Pixiu is a combined creature that looks slightly particular relying on whilst it have become made and wherein it comes from. Most of the time, even though, the pixiu's body looks as if a cat, a dog, or a dragon. The pixiu can also have wings as well as a head now and again. Plus, the pixiu is normally tested as a monster that sits on its hind legs like a canine, which makes it look very strong. It could probable even appearance a touch fat.

Remember that the pixie is seen as a fortunate animal in Chinese folklore. The dragon, the ocean turtle, Kylin, and the phoenix are all within the same business enterprise. Because the Pixiu is this type of sturdy and threatening animal, it in fact works to be a father or mother of Heaven, and its number one gadget is to fight off ghosts and devils. And just like the Kylin and the dragon, a Pixiu is fairly actual because it makes the humans that possess it glad and additionally can be used to eliminate evil spirits. Unlike the Kylin, the Pixiu is a unstable animal that protects its Master with all its strength. This is why many Chinese people placed jade Pixius on their our bodies. It's moreover crucial to realize that the Pixiu has 40 nine specific paperwork and 26 figures.

Well, unique Chinese memories tell one-of-a-kind reminiscences approximately this fortunate creature. Most of the time, even though, human beings speak approximately the pixiu because the youngest of the Dragon King's nine youngsters. The Dragon King is each exceptional legendary creature. The Dragon King and dragons are crucial for Chinese people due to the truth they're concept to have magical powers. People anticipate dragons can trade the climate, rule over the seas, and are very strong. This way that their sons are also effective. The dragon is a powerful image of honour and power in Chinese society.

The tale goes that one time, the Pixiu went to appearance the Jade Emperor, however ended up going to the rest room anywhere in the palace floors. The Pixiu's actions made the Jade Emperor angry, so he took the extreme step of sealing off the Pixiu's complete backside. This ought to stop the Pixiu from pooping ever all all over again, so mistakes like this will in no way seem yet again.

Myths moreover describe what the Pixiu did and preferred. For example, one story says that the Pixiu favored to devour costly such things as gold and silver jewelry. Because the Pixiu's bottom modified into sealed off, these valuables ought to in no way leave its frame. Because of this, it have emerge as a signal of wealth and loads as it ate treasures however ought to by no means get rid of them. In distinct phrases, humans belief the Pixiu might also want to get hold of wealth and make it take area for its owner.

Experts in Feng Shui say that the Pixiu keeps houses secure. Getting rid of evil spirits and demons from homes is what it does. It moreover brings top fortune to its proprietor. It is said in Feng Shui that in case you contact the Pixiu 3 instances, you could get more authority and the excessive function you need. Touch it as quickly as for suitable success, instances for cash and first rate wealth, and three times for greater power and the vicinity you need.

The Pixiu is carved from 3 primary materials: Jade, copper, and wooden. The copper gives it a gold-like end after it is been polished, but the timber became to start with used to make the Pixiu. Jade has been used to carve the maximum up to date Pixiu styles. The Pixiu figures frequently take a seat down down on pinnacle of big homes. It became idea that this may preserve terrible achievement away. You could be satisfied to recognize that the Pixiu normally has an equal quantity of strength, no matter what substances have been used to make it, so long as you hold it inside the right place.

Chapter 6: Pigman

The contemporary-day media display Zhu Bajie as a lovely and notable-searching red pigman, which has induced plenty controversy approximately his look. But the ebook makes Zhu proper right into a big pig monster with a protracted head, large ears together with fans, and a massive mouth. Because Zhu's body has components of every human beings and animals, he has humanoid arms, toes, and a pig's tail. He is awesome and has frightening black fur on his pores and pores and pores and skin.

According to ancient assets, Zhu Bajie changed into one among Tang Sanzang's 3 helpers. He is also a widespread person in "Journey to the West." In the e-book, he's confirmed to be an disturbing and lazy character who constantly receives others into hassle. His naughtiness is confirmed via way of the fact that he's in reality into ladies, can not prevent consuming, and refuses to labour hard. He moreover comes at some stage in as very jealous, confirmed nicely thru how difficult he works to hold down Sun Wukong.

Zhu Bajie became portrayed as a effective individual. In the story, he persuaded the older to allow his daughter marry him through saying that he will be able to do masses artwork because of the fact he changed into so robust. However, the request is have emerge as down even as evidently Zhu Bajie is able to tons crop labour but eats lots on the farm, this means that that that he loses coins as opposed to making it.

In his past existence, Zhu Bajie lived within the heavens. He become a well-known individual with the name Tianpeng Yuanshuai and have come to be in price of as lots as eighty,000 Heaven Navy Soldiers. However, he didn't stay in this process for prolonged due to the fact he have become fired for being incorrect. History says that

Zhu Bajie attended a celebration for all of the critical people in heaven.

When Bajie noticed the deity of the night time time for the first time on the birthday celebration, he became blown away thru how stunning she changed into. After that, he tried to seduce her on the same time as underneath the have an impact on of alcohol, but it did not artwork, which turn out to be terrible for him. The moon goddess informed the Jade Emperor about Zhu's behaviour, which precipitated Zhu to be despatched decrease lower back to Earth.

Some books say that Zhu become sent to Earth and will die one thousand instances. But every one of the a thousand memories can also need to come to an result in a sad love tale. When he have grow to be despatched to Earth, he fell right into a pig well via manner of mistake, which is how he had been given his pig-like competencies. So, Zhu lower back to life as a pig monster that ate people. Then, it changed into given the decision Zhu Ganglie, because of this "sturdy-armed pig" in English. As destiny can also have it, Zhu Bajie grieved because of the fact he cherished the sturdy-maned pig, which took the elder's daughter and left a be aware inquiring for marriage.

Chang'e worked for a showgirl within the celestial palace, and the tale goes that Zhu Bajie permit his morals slip on the equal time as he turned into beneath the affect of alcohol. He went on a liaison with Chang'e, and he saved pushing her to relaxation with him. Then, human beings suppose that Zhu Bajie raped Chang'e. It is perception that little has occurred among Zhu and Chang'e in different factors. When it came out that Zhu Bajie had abused Chang'e, the ruler Jade killed him after eating and sent him decrease lower back to the real global. Zhu come to be sorted thru using the woman who have been his second sister egg inside the worldwide.

While touring, demons may change into children or ladies who favored help. This have become finished to trick them and get Master Tang Sanzang to devour them. But Sun Wukong may additionally want to sense demons. On the opposite hand, Zhu talked his brother into letting them circulate in place of catching and killing them. Zhu's kindness did no longer get him anywhere; it regularly were given him into problems. Zhu regarded as much as and revered Sun Wukong and brought into attention him his brother. On the opportunity hand, Sun in no way thought masses of Zhu; instead, he made amusing of him and modified into commonly at odds with him. People think that Zhu only proper Sun Wukong due to the fact that he knew that Sun Wukong have become a remarkable boxer in a past existence.

Demons are continuously taking Zhu on the identical time as he is at the journey. However, this does not trouble him due to the fact he commonly acts although the spirits need to consume him. His quality mind-set comes from in advance than he become born, while he learnt a way to cope with many terrible subjects. Zhu found out a manner to stay on pinnacle of things of his emotions. Some components of the tale say that Zhu did not want to apply their natural capabilities at the journey. He idea the monkey king could shop him and the others, and if that did now not take area, the alternative gods might also.

People within the ebook "Journey to the West" recollect Zhu Bajie as a repulsive individual who continuously receives into hassle. He is robust, but he could now not want to utilize his abilities. This is seen as lazy on his element due to the fact he thinks the opportunity gods will preserve him. Even despite the fact that he seems like a slave to love, he is sent from the clouds whenever he tries to grow to be near a person he finds stunning.

Chapter 7: China

Chinese dragon myths are an imperative part of Chinese folklore, artwork, and literature. These myths date again masses of years and have performed a huge function in shaping Chinese manner of existence and facts. Here are some key factors of Chinese dragon myths:

Appearance: Chinese dragons are commonly depicted as lengthy, serpentine creatures with 4 legs, sharp claws, and a horned head. They are blanketed in scales and characteristic a snake-like tongue. Unlike Western dragons, Chinese dragons are not depicted as respiratory fireplace.

Symbolism: Chinese dragons are taken into consideration to be effective and benevolent creatures, associated with notable achievement, prosperity, and strength. They also are seen as symbols of imperial electricity and are regularly depicted in paintings and literature as protectors of the emperor.

Origins: The origins of Chinese dragon myths aren't absolutely easy, but they will be believed to have been stimulated thru diverse creatures from Chinese mythology, together with the qilin (a creature with the top of a dragon and the frame of a deer) and the nian (a monster that grow to be scared off by means of manner of firecrackers inside the direction of the Chinese New Year).

Cultural Significance: Chinese dragon myths have completed an vital position in Chinese way of life, with dragon dances and dragon boat festivals being famous activities ultimately of China and the area. Chinese dragons also are featured in numerous works of art and literature, together with the traditional Chinese novel Journey to the West.

Overall, Chinese dragon myths are a charming and critical part of Chinese manner of lifestyles and statistics, and maintain to captivate human beings round the arena these days.

The Black Dragon

In Chinese mythology, the black dragon is called Hei Long, which really method "black dragon." It is one of the 4 symbols of the Chinese constellations and is associated with the course north and the element water.

According to Chinese myth, the black dragon is the ruler of the underworld and is often associated with loss of existence and destruction. It is stated to be a fierce and effective creature, capable of inflicting earthquakes and special natural screw ups.

In Chinese artwork, the black dragon is regularly depicted as a protracted, serpentine creature with a black body and glowing red eyes. It is every now and then shown carrying a pearl or a flaming sword, symbols of its energy and authority.

Despite its association with loss of life and destruction, the black dragon is also seen as a image of rebirth and renewal. In a few myths, it's miles said to emerge from the depths of the earth to deliver new life and electricity to the arena.

Overall, the black dragon is a complicated and effective image in Chinese mythology, representing both the unfavorable and innovative forces of nature. Its delusion maintains to captivate and inspire people in China and spherical the arena nowadays.

The Yellow Dragon

In Chinese mythology, the yellow dragon is known as Huang Long, which actually way "yellow dragon". It is one of the 4 symbols of the Chinese constellations and is related to the course center and the element earth.

According to legend, the yellow dragon grow to be born from the mom of the number one emperor of China. The emperor's mom dreamt that a divine dragon impregnated her, and she or he or he gave delivery to the yellow dragon, who have become said to have an extended, serpentine frame with yellow scales and a couple of antlers on its head.

The yellow dragon is taken into consideration to be a photograph of imperial electricity and is often associated with the Chinese emperor. It is concept to have the electricity to control the waters and is sometimes depicted as retaining a pearl or a e-book in its claws, representing expertise and focus.

In Chinese life-style, the yellow dragon is also related to properly fortune and prosperity. It is regularly depicted in paintings and architecture, together with at the roofs of temples and palaces.

Overall, the yellow dragon is a powerful and revered image in Chinese mythology and stays an vital part of Chinese life-style these days.

The Pearl Dragon

In Chinese mythology, the Pearl Dragon is known as Zhu Long, which genuinely way "pearl dragon". The Pearl Dragon is a benevolent and non violent creature this is associated with the moon and the night time time sky.

According to legend, the Pearl Dragon end up born from a pearl that have end up customary inside the mouth of a large clam. The pearl modified into said to comprise the essence of the moon and changed into guarded with the aid of the usage of the clam. The dragon emerged from the pearl and have become tasked with defensive it.

The Pearl Dragon is often depicted as having a frame included in shimmering pearls or retaining a pearl in its claws. It is likewise sometimes set up riding on clouds or inside the corporation of diverse celestial creatures, which include phoenixes and unicorns.

In Chinese way of life, the Pearl Dragon is taken into consideration to be a image of attention, purity, and correct fortune. It is often depicted in art work and jewellery, which incorporates on jade carvings and pearl necklaces.

Overall, the Pearl Dragon is a peaceful and respected photo in Chinese mythology, representing the beauty and electricity of the natural worldwide. Its delusion maintains to encourage and captivate people in China and spherical the arena in recent times.

The Dragon Kings

In Chinese mythology, the Dragon Kings are a hard and fast of effective dragons who're said to rule over the seas and control the weather. They are believed to be the sons of Longwang, the dragon king of the ocean, and are related to the four seas of China: the East Sea, the South Sea, the West Sea, and the North Sea.

According to legend, every Dragon King policies over a specific sea and has the power to govern the tides and the weather. They are stated to live in underwater palaces and are followed thru a retinue of aquatic creatures, inclusive of fish, turtles, and crabs.

The Dragon Kings are also associated with the Chinese calendar, with every dragon representing a extraordinary season. The Azure Dragon, for example, is related to spring, while the Black Dragon is related to wintry weather.

In Chinese way of life, the Dragon Kings are taken into consideration to be powerful and benevolent creatures, related to right exact fortune, prosperity, and the herbal international. They are regularly depicted in artwork and literature, which encompass within the traditional Chinese novel Journey to the West, and are featured in severa fairs and ceremonies during China.

There are 4 Dragon Kings who every rule over a awesome sea. They are:

Ao Guang - He is the Dragon King of the East Sea and is associated with the spring season. He is regularly depicted with a dragon's head and a human body and is stated to control the tides and the rain.

Ao Qin - He is the Dragon King of the South Sea and is associated with the summer season. He is regularly depicted with a dragon's head and a human frame and is said to govern the winds and the clouds.

Ao Run - He is the Dragon King of the West Sea and is related to the fall season. He is frequently depicted with a dragon's head and a human frame and is said to govern the rivers and the lakes.

Ao Shun - He is the Dragon King of the North Sea and is related to the wintry weather season. He is regularly depicted with a dragon's head and a human frame and is said to govern the bloodless and the ice.

These Dragon Kings are believed to be effective and benevolent creatures who defend and govern the seas and the herbal international. They are an critical part of Chinese mythology and continue to be celebrated and revered in Chinese way of life nowadays.

Ao Guang

The Dragon King Ao Guang is a character from Chinese mythology and folklore. He is referred to as the ruler of the Eastern Seas and is considered one of the most powerful and revered of the dragon kings.

According to legend, Ao Guang lived in a surprising crystal palace at the bottom of the sea. He became seemed for his capability to govern the tides and the rain, and became reputable thru fishermen and sailors who believed that he have to defend them from the dangers of the sea.

One day, the famous Monkey King, Sun Wukong, brought on chaos in the Dragon

Palace through manner of stealing the prized weapon of Ao Guang, a mystical staff that could alternate its duration at will. The Dragon King become livid and ordered his military of sea creatures to capture Sun Wukong.

However, the Monkey King proved to be too powerful for the dragon king's military and defeated all of them. Impressed thru manner of Sun Wukong's power and bravery, Ao Guang decided to make a address him. He furnished to offer him the place of "Keeper of the Heavenly Horses" in exchange for the body of workers.

Sun Wukong agreed to the deal, however he tricked the Dragon King by the usage of the usage of transforming himself right proper right into a tiny flea and hiding on the tip of the frame of personnel. When Ao Guang handed over the body of workers, Sun Wukong transformed decrease returned to his actual shape and defeated the dragon king's army over again.

Although the Dragon King have come to be defeated, he professional Sun Wukong's strength and gave him the region he had promised. From then on, the Monkey King and the Dragon King have turn out to be buddies and allies, walking together to defend the seas and the heavens from threat.

Ao Qin

The Dragon King Ao Qin is a individual from Chinese mythology and folklore. He is called the ruler of the Western Seas and is taken into consideration one of the most effective and revered of the dragon kings.

According to legend, Ao Qin lived in a wonderful palace made from pink and white coral at the lowest of the sea. He grow to be referred to for his capacity to govern the winds and the clouds, and became revered via sailors and fishermen who believed that he need to shield them from storms and typhoons.

One day, a more youthful pupil named Liu Yi became travelling during the Western Seas on the equal time as he encountered a powerful hurricane. He prayed to the gods for help, and Ao Qin heard his prayers and got here to his resource.

Impressed via Liu Yi's information and know-how, Ao Qin invited him to his palace and asked him to live as his tourist. Liu Yi changed into amazed through using the wonders of the Dragon King's palace, and he soon have emerge as buddies with Ao Qin and his family.

During his live, Liu Yi found that the Dragon King's state turned into tormented by a drought. He provided to assist via way of bringing rain to the Western Seas, but Ao Qin warned him that first-class the gods had the strength to manipulate the rain.

Undeterred, Liu Yi studied the ideas of climate and devised a plan to deliver rain to the drought-bothered u . S .. He created a tool that might manage the clouds and the winds, and used it to supply rain to the Western Seas.

Impressed by using way of Liu Yi's capability and understanding, Ao Qin made him the leader advisor of his united states. From then on, Liu Yi worked with the Dragon King to defend the seas and the people of the Western Seas from harm.

Ao Run

The Dragon King Ao Run is a individual from Chinese mythology and folklore. He is referred to as the ruler of the Northern Seas and is considered one of the maximum effective and revered of the dragon kings.

According to legend, Ao Run lived in a great palace made of ice and pearls at the lowest of the ocean. He was seemed for his capability to govern the snow and the ice, and was respected with the aid of the those who lived inside the a ways north.

One day, a set of tourists were lost in a blizzard and stumbled upon Ao Run's palace. The Dragon King welcomed them and furnished them steady haven from the typhoon. Impressed with the aid of his kindness, the vacationers provided to help him in any manner they will.

Ao Run defined that his nation changed into tormented by a terrible drought, and he feared that his human beings might starve. The tourists provided to assist via the use of finding a way to supply water to the Northern Seas.

They set out on a long journey and in the end came across a mystical fountain that had the electricity to create rivers and lakes. They introduced the fountain back to Ao Run's palace, and the Dragon King used it to hold water to his united states.

Impressed with the beneficial useful resource of their bravery and ingenuity, Ao Run made the travelers his advisors and welcomed them to his united states of the united states. From then on, the Dragon King and his new advisors labored together to protect the Northern Seas and ensure that his people never went thirsty once more.

Ao Shun

The Dragon King Ao Shun is a character from Chinese mythology and folklore. He is known as the ruler of the Southern Seas and is taken into consideration one of the satisfactory and revered of the dragon kings.

According to legend, Ao Shun lived in a remarkable palace made from red and inexperienced jade at the bottom of the ocean. He have turn out to be stated for his capacity to control the rain and the mist, and modified into revered with the aid of the use of way of the people who lived within the southern regions.

One day, a collection of fishermen had been caught in a terrible typhoon, and their boat emerge as destroyed. They prayed to the gods for assist, and Ao Shun heard their prayers and came to their useful resource.

He rescued the fishermen and taken them to his palace, where he furnished them meals and safe haven. Impressed with the resource of his kindness, the fishermen supplied to help him in any manner they could.

Ao Shun defined that his united states modified into stricken by a horrible drought, and he feared that his people might die of thirst. The fishermen furnished to help via locating a way to keep water to the Southern Seas.

They set out on an prolonged journey and in the long run came throughout a magical pearl that had the electricity to create rain and dew. They introduced the pearl once more to Ao Shun's palace, and the Dragon King used it to bring water to his kingdom.

Impressed through their bravery and ingenuity, Ao Shun made the fishermen his advisors and welcomed them to his u . S . A .. From then on, the Dragon King and his new advisors worked collectively to shield the Southern Seas and make certain that his humans in no manner went thirsty again.

Chapter 8: Germany

German mythology has numerous reminiscences and legends approximately dragons, a number of which might be despite the fact that famous in German folklore these days. Here are some examples:

Fafnir: In Norse mythology, Fafnir become a dragon who guarded a first rate treasure hoard. According to the legend, Fafnir became as quickly as a dwarf who have turn out to be transformed proper right into a dragon via manner of a cursed ring. He was in the long run slain thru the hero Sigurd (furthermore called Siegfried), who used his sword to pierce Fafnir's coronary heart.

Lindworm: The Lindworm is a dragon-like creature in Germanic mythology that is typically depicted with legs in area of 4. It is notion for its fierce temper and its potential to breathe fireside. In a few stories, the Lindworm is portrayed as a mother or father of a treasure hoard, on the equal time as in others it's far a chance to the local people.

Drachenfels: The Drachenfels is a mountain positioned near Bonn, Germany, that is named after a dragon. According to legend, the dragon lived in a cave at the mountain and terrorized the close by population till it turn out to be slain thru Siegfried. The Drachenfels is now a well-known traveler tour spot, and site visitors can climb to the top of the mountain to appearance the ruins of a fortress that have become built there within the twelfth century.

These are just a few examples of the numerous dragon myths and legends which might be a part of German folklore. Like many myths and legends, they offer insights into the values, ideals, and fears of the people who created them.

Fafnir

The myth of Fafnir is part of Norse mythology and is likewise famous in German folklore. Fafnir become a dragon who guarded a splendid treasure hoard, and he emerge as appeared for his first-rate period, fierce temper, and capability to respire hearth.

According to the legend, Fafnir changed into as quickly as a dwarf who modified into converted right proper into a dragon by way of of the usage of a cursed ring. He and his brother, Regin, were each professional blacksmiths who had inherited a remarkable treasure hoard from their father. However, Regin have come to be fed on with greed and convinced Fafnir to kill their father and take the treasure for themselves.

After killing their father, Fafnir have turn out to be paranoid and retreated to a far flung mountain cave to protect the treasure. He converted proper right into a dragon to protect himself and the treasure, and he have come to be an increasing number of opposed to absolutely everyone who dared to return close to.

The hero Sigurd, furthermore known as Siegfried, have become subsequently sent to slay Fafnir and claim the treasure. He changed into aided via Regin, who had happy him to adopt the venture within the first vicinity. Regin provided Sigurd with a sword capable of killing the dragon, however he moreover secretly plotted to betray him and claim the treasure for himself.

Sigurd slew Fafnir with the sword, but he also acquired the capability to apprehend the language of birds after eating the dragon's coronary coronary heart. The birds warned Sigurd of Regin's treachery, and he in the end killed Regin as nicely earlier than claiming the treasure for himself.

The story of Fafnir is often seen as a cautionary story about the risks of greed and the bad strength of dragons. It is likewise a well-known problem in German folklore and has been tailored into plays, operas, and unique works of artwork over the centuries.

Lindworm

The Lindworm is a dragon-like creature in Germanic mythology this is normally depicted with legs rather than 4. It is understood for its fierce mood and its functionality to respire fireside. The delusion of the Lindworm varies specifically factors of Germany, but a famous model is as follows:

Once upon a time, a queen changed into searching earlier to a little one, however while the kid changed into born, it became a Lindworm in area of a human toddler. The queen modified into afraid and ordered the Lindworm to be taken away and deserted within the wooded region. The Lindworm grew up by myself inside the forest, and because it grew, it have come to be an increasing number of fierce, causing destruction and terrorizing the close by villages.

One day, a prince heard approximately the Lindworm and determined to slay it to show his bravery. He traveled to the wooded location and discovered the Lindworm, however as he attempted to kill it, he changed into surprised to locate that it can speak. The Lindworm begged the prince to spare its life and promised to prevent inflicting harm to the villagers if the prince may spare its existence.

The prince agreed to spare the Lindworm's existence at the situation that it leave the village and in no manner flow once more. The Lindworm agreed and left the vicinity, however it decrease decrease back years later, having grown even big and more effective. This time, the prince end up now not able to defeat it, and he turn out to be pressured to are looking for the help of a sensible vintage woman.

The smart antique female gave the prince advice on the way to defeat the Lindworm. She knowledgeable him to use a defend made from linden wooden, which the Lindworm is probably no longer capable of penetrate. The prince discovered the recommendation and modified into capable of slay the Lindworm, saving the village from its terror.

The tale of the Lindworm is often visible as a caution in competition to the dangers of forsaking or mistreating kids or creatures, further to a reminder of the energy of bravery and interest in the face of hazard. The Lindworm is a famous creature in German folklore and has been tailored into performs, operas, and special works of paintings over the centuries.

Drachenfels

Drachenfels is a mountain in Germany this is surrounded by way of fantasy and legend. In Germanic mythology, it's miles believed to be the residence of dragons, which have been said to guard a extraordinary treasure hoard hidden in the mountain.

One famous legend tells of a knight named Siegfried who slayed the dragon that lived on Drachenfels and claimed its treasure. According to the legend, Siegfried bathed in the blood of a dragon and have become invincible. He then went on to slay the dragon that lived on Drachenfels and declare its treasure, which blanketed a paranormal ring that gave him notable strength.

Another model of the legend tells of a prince who fell in love with a beautiful maiden who turned into imprisoned in a fortress on pinnacle of Drachenfels. The citadel have become guarded by using using way of a dragon, and the prince became unable to rescue the maiden until he positioned a way to slay the dragon.

Today, Drachenfels is a well-known vacationer enchantment and a symbol of Germanic mythology and folklore. Visitors can hike to the pinnacle of the mountain to see the fort ruins and take inside the lovely perspectives of the Rhine River valley.

Nidhogg

Nidhogg is a distinguished determine in Germanic mythology, and is depicted as a dragon or serpent who gnaws at the roots of the sector tree, Yggdrasil. The name "Nidhogg" approach "tearer of corpses" or "striker," and it is stated that he feeds on the corpses of the vain.

According to legend, Nidhogg is living at the lowest of Yggdrasil, in a region known as Niflheim, the arena of ice and darkness. Nidhogg's normal gnawing on the roots of Yggdrasil is stated to purpose the tree to wither and decay, and plenty of trust that Nidhogg's ultimate goal is to result in the forestall of the area.

Despite his detrimental nature, Nidhogg is likewise believed to serve an critical role in keeping the stableness of the cosmos. The worldwide tree, Yggdrasil, is stated to connect the 9 worlds of Germanic mythology, and is a picture of the interconnectedness of all matters. Nidhogg's gnawing on the roots of the tree is seen as a herbal technique of decay and renewal, which permits new lifestyles to develop and flourish.

In a few interpretations of the Nidhogg fantasy, the dragon is visible as a photo of greed and avarice, preying at the corpses of the useless to fulfill his insatiable starvation. In others, he's visible as a effective stress of nature, embodying the unfavorable power of the elements.

Despite the numerous versions within the Nidhogg fable, the dragon stays a powerful and enduring figure in Germanic mythology. His feature as a photograph of destruction, renewal, and balance makes him an important a part of the complex tapestry of Germanic folklore and mythology.

Britain

There are several dragon myths and legends related to British folklore, and that they variety throughout unique areas and time periods. Here are a number of the most tremendous ones:

The Red Dragon of Wales: This is probable the most well-known dragon fable associated with Britain. According to Welsh legend, the Red Dragon become the logo of the Welsh king Cadwaladr, who fought towards the invading Saxons. The dragon is stated to have defeated the White Dragon of the Saxons, therefore securing Wales as an unbiased united states.

The Lambton Worm: This is a legend from the northeast of England, which tells the story of John Lambton, a greater youthful man who catches a weird creature while fishing inside the River Wear. He throws the creature away, however it grows right right into a large pc virus that terrorizes the place. Eventually, Lambton returns to the river and enlists the help of a clever female, who tells him to coat his armor in spearheads and confront the trojan horse inside the river. He does so, and after a fierce war, the computer virus is defeated.

The Dragon of Wantley: This is a funny legend from Yorkshire, which tells the story of a dragon that terrorizes the geographical region with the aid of using the use of ingesting farm animals or even villagers. The knight More of More Hall is referred to as upon to defeat the dragon, and he does so through way of sporting a in shape of spiked armor and attacking the dragon's inclined spots.

The Knucker: This is a dragon from Sussex, which end up said to stay in a pool close to the village of Lyminster. The Knucker became said to be a fierce creature that demanded a sacrifice of a milkmaid each twelve months. Eventually, a close-by hero named Jim Puttock managed to defeat the Knucker thru tricking it into eating a toxic pie.

These are only some examples of the dragon myths and legends related to British folklore. Dragons have been a well-known a part of British folklore for masses of years, and they preserve to capture the creativeness of people nowadays

Chapter 9: Red Dragon Of Wales

The fable of the Red Dragon of Wales is a tale from Welsh mythology that tells the tale of two dragons - one crimson and one white - that fight every unique for supremacy.

According to the parable, the crimson dragon represents the people of Wales, on the same time due to the fact the white dragon represents the invading Saxons. The dragons are said to had been buried in a hill in Snowdonia, a mountain range in Wales.

The tale goes that King Vortigern, a 5th-century ruler of the Britons, become on the lookout for to collect a fort at the hill however the walls saved collapsing. His advisors encouraged him that the only manner to resolve the hassle has become to find out a boy without a father and sacrifice him on the internet website on-line. Vortigern observed this sort of boy named Merlin, however Merlin recommended him that the real reason for the crumble emerge as that there were dragons drowsing under the hill.

Merlin informed Vortigern to dig up the hill and the 2 dragons emerged, one purple and one white. The dragons fought fiercely and ultimately, the crimson dragon emerged effective. Merlin explained that the red dragon represented the Welsh people and their victory over the invading Saxons.

The fantasy of the Red Dragon of Wales has grow to be an important image of Welsh identification and nationalism. The flag of Wales competencies a purple dragon on a inexperienced and white history, and the story has been celebrated in Welsh literature and artwork for loads of years.

Lambton malicious program

The Lambton Worm is a legendary creature from northern England, specially the county of Durham. The delusion tells the tale of a younger man named John Lambton, who catches a regular, wriggling creature on the identical time as fishing in the River Wear.

Not understanding what it's far, John tosses the creature into a nearby properly, in which it keeps to increase and finally turns into the Lambton Worm, a big, serpentine monster with sharp teeth and sparkling eyes.

The Lambton Worm terrorizes the geographical area, killing livestock and eating villagers. John Lambton, wracked with guilt over what he has unleashed, consults with a smart lady who tells him that the pleasant manner to defeat the monster is to make a match of armor covered in sharp blades and confront it in the river.

John does as he is usually recommended, and after a fierce war, he's able to slice the Lambton Worm into quantities. However, the clever girl warns John that he should in no manner boast approximately his victory, otherwise his circle of relatives will undergo a curse for nine generations.

John returns domestic to locate that his father has died and that the curse has taken maintain. However, after nine generations have exceeded, a member of the Lambton family joins the Crusades and returns with a everyday animal that he indicates off in a public show. The clever lady, now an vintage lady, acknowledges the creature because the Lambton Worm and curses the own family all over again, saying that their property will stay barren till a family member redeems their ancestor's sin.

The legend of the Lambton Worm has emerge as an essential a part of folklore within the north of England, and the tale has been retold in numerous forms, collectively with ballads, performs, and novels.

The Dragon of Wantley

The Dragon of Wantley is a medieval English folks story that tells the tale of a dragon terrorizing the village of Wantley in Yorkshire.

According to the myth, the dragon come to be a fierce beast with sharp enamel, great wings, and an insatiable urge for food for cattle and villagers. The people of Wantley tried severa methods to defeat the dragon, however all tries failed, and lots of brave knights had been killed within the method.

The villagers ultimately grew to grow to be to a knight named More of More Hall, who've become stated for his strength and bravery. More of More Hall devised a foxy plan to defeat the dragon, which worried eating a special diet plan of hedgehogs, after which sporting armor made from spikes and sharp edges.

When the dragon attacked, More of More Hall fought it fiercely, and the spikes on his armor caused the dragon to bleed and weaken. More of More Hall then captured the weakened dragon, and with the assist of the villagers, killed it.

The legend of the Dragon of Wantley has end up an critical a part of English folklore, and the story has been retold in numerous office paintings, consisting of ballads, plays, and novels. The tale has also inspired inventive depictions, which includes William Hogarth's portray "The Yorkshire Dragon."

Knucker

The Knucker is a mythical dragon-like creature from Sussex, England. The delusion tells the story of a dragon that lived in a deep pool of water known as the Knucker Hole.

According to the legend, the Knucker terrorized the nearby villagers, consuming their livestock or even on occasion snatching up kids. The human beings of Sussex were scared of the Knucker and couldn't discern out the way to defeat it.

One day, a more youthful guy named Jim Puttock determined to take topics into his very own arms. Jim had heard that the Knucker changed right right into a vain creature that enjoyed to reveal off its electricity, so he made a plan to trick the Knucker.

Jim dug a big pit close to the Knucker Hole and guarded it with branches and leaves to make it appear to be a natural a part of the landscape. He then challenged the Knucker to return again and fight him, boasting that he was now not terrified of the creature.

The Knucker took the bait and came out of its hole to fight Jim. As they battled, Jim tricked the Knucker into falling into the pit, in which it have come to be caught. Jim then killed the Knucker, turning into a hero inside the eyes of the villagers.

The myth of the Knucker has end up an crucial a part of Sussex folklore, and the tale has been retold in severa office work, which encompass ballads, plays, and novels. The Knucker is also featured in close by artwork and fairs, which include the Knucker Fair in Lancing.

Mesopotamia

Mesopotamian dragon myths are a fascinating part of historical Mesopotamian mythology, which grow to be one of the earliest recorded mythologies within the international. Mesopotamian dragon myths feature diverse kinds of dragons, each with its precise developments and competencies.

One of the most well-known Mesopotamian dragons changed into Tiamat, who changed into stated to be the mom of all dragons. In Mesopotamian mythology, Tiamat have end up depicted as a big, multi-headed dragon that lived within the depths of the sea. Tiamat changed into believed to be considerably powerful and became stated a superb manner to manipulate the forces of chaos.

Another well-known Mesopotamian dragon became Mushussu, a dragon with the pinnacle of a serpent and the forelegs of a lion. Mushussu became related to the god Marduk, who come to be believed to have defeated the dragon and used its photograph as a symbol of his electricity.

Other Mesopotamian dragons embody Asag, who was stated to were born from the rocks and changed right right into a fierce enemy of the gods, and Sirrush, a dragon-like creature with the top of a lion and the frame of a serpent that modified into related to the goddess Ishtar.

Mesopotamian dragon myths frequently had the dragon as a image of chaos and destruction, and the defeat of a dragon become visible as a victory of order over chaos. Dragons had been moreover related to the powers of the underworld and had been believed to be guardians of treasure and sacred information.

Overall, Mesopotamian dragon myths provide a captivating glimpse into the historic beliefs and cosmology of one of the international's earliest civilizations.

Tiamat

The fantasy of Tiamat is a number one tale in Mesopotamian mythology, which grow to be one of the earliest recorded mythologies in the global. According to the parable, Tiamat modified into the primordial goddess of the sea, who gave start to the gods and changed into seen because of the fact the mom of all advent.

However, through the years, the gods started out out to develop restless and conspired in competition to Tiamat, plotting to overthrow her and take manipulate of the universe. Led thru the usage of the god Marduk, they waged a splendid conflict in opposition to Tiamat and her big navy, which protected numerous varieties of dragons and other fearsome creatures.

In the warfare, Marduk confronted Tiamat and used his magical powers to interrupt up her in , growing the heavens and the earth from her frame. From Tiamat's blood, Marduk created human beings, and he set up the hierarchy of the gods, with himself because the very best ruler.

After the war, Marduk end up praised by using using the opportunity gods, who declared him to be the brilliant of all the gods. They constructed a first-rate temple in his honor, and he have turn out to be the maximum crucial deity within the Mesopotamian pantheon.

The tale of Tiamat and her defeat by using way of Marduk is frequently visible as a metaphor for the triumph of order over chaos, and it performed a terrific function in Mesopotamian spiritual and political life. The story also suggests the significance of mythology in shaping ancient societies' beliefs and values, in addition to their statistics of the universe and their location in it.

Mushussu

In Mesopotamian mythology, Mushussu modified proper into a dragon-like creature with the pinnacle of a serpent and the forelegs of a lion. The delusion of Mushussu is cautiously tied to the tale of the god Marduk, who became believed to have defeated the creature and used its photograph as a photo of his electricity.

According to the myth, the god Marduk become tasked with defeating a powerful demon named Tiamat, who had created an military of monsters to combat in opposition to the gods. One of these monsters changed into Mushussu, who changed into said to had been made from the spittle of the demon Kingu.

Marduk fought bravely in opposition to Tiamat and her military of monsters, and he in the end defeated them thru growing a effective wind that swept them away. As a praise for his victory, the opportunity gods declared Marduk to be the very super ruler of the universe and gave him the find out of "King of the Gods."

In honor of his victory over the monsters, Marduk followed the photograph of Mushussu as his personal photograph. The picture of Mushussu end up often depicted on seals and one-of-a-kind devices associated with the worship of Marduk, and it have grow to be a sturdy image of the god's power and authority.

In Mesopotamian mythology, Mushussu grow to be moreover occasionally related to the goddess Ishtar, who became believed to have used the creature as her mount or steed. The photograph of Mushussu as a powerful, dragon-like creature become one of the most enduring symbols of Mesopotamian mythology, and it persevered for use in art and spiritual iconography long after the civilization itself had exceeded into statistics.

Asag

In Mesopotamian mythology, Asag have come to be a effective demon who have become believed to had been born from the rocks of the earth. He modified into depicted as a fearsome creature with the body of a serpent, the legs of a donkey, and the pinnacle of a lion or a canine.

According to the myth of Asag, the demon changed right into a fierce enemy of the gods and feature emerge as liable for bringing chaos and destruction to the world. He changed into stated to have created terrific storms and prompted earthquakes and one-of-a-type natural failures.

In reaction to Asag's rampage, the gods despatched the hero Ninurta to defeat him. Ninurta modified into a effective warrior god who come to be recognized for his bravery and his skills in conflict. He faced

Asag in a extraordinary showdown, the usage of his magical weapons and his information of the earth to defeat the demon.

In the give up, Ninurta have turn out to be triumphant, and Asag changed into banished to the underworld, wherein he could not threaten the world along along with his chaos and destruction. The defeat of Asag grow to be seen as a triumph of order over chaos, and it helped to installation Ninurta as one of the maximum critical gods within the Mesopotamian pantheon.

The fable of Asag is huge as it suggests the significance of mythology in Mesopotamian society. The story helped to offer an motive at the back of natural phenomena together with earthquakes and storms, and it also bolstered the concept that the gods had been on pinnacle of things of the universe and could intervene to protect humanity from evil forces.

Sirrush

In Mesopotamian mythology, Sirrush (moreover called Mushhushu) became a mythical creature that resembled a dragon or a serpent with the hind legs of a hen of prey. It come to be related to the god Marduk and became taken into consideration a image of his energy.

According to the myth, Sirrush changed into created through the usage of the use of Tiamat, the primordial goddess of the ocean, as one of the large creatures to oppose the gods. When Marduk defeated Tiamat and her navy of monsters, he claimed Sirrush as his very non-public.

Marduk then used the image of Sirrush as his personal image, and it modified into frequently depicted on devices related to his worship, together with seals, amulets, and extremely good gadgets of strength. Sirrush grow to be moreover believed to have served as a mount for Marduk, wearing him across the sky as he battled the forces of chaos.

The delusion of Sirrush is big because it demonstrates the significance of symbolic example in Mesopotamian subculture. The creature have become seen as a photo of Marduk's strength and authority, and its photo become used to enhance the idea that Marduk end up the ideally fitted god in the Mesopotamian pantheon.

Sirrush furthermore finished a role in Mesopotamian artwork and iconography, and plenty of examples of its image were decided in archaeological excavations. Today, the picture of Sirrush stays diagnosed as a effective picture of Mesopotamian mythology and the historic cultures that superior in the area.

Arabia

Arabian dragons are legendary creatures from the folklore and mythology of the Arabian Peninsula. They are often depicted as serpentine or reptilian creatures with wings and the capability to respire fireplace or special elemental powers. Arabian dragons also are normally related to wealth and treasure, and are often depicted as guardians of valuable devices or places.

One well-known Arabian dragon is the Zahhak, additionally referred to as the Dragon King. In Persian mythology, Zahhak emerge as a cruel tyrant who modified into cursed with serpents developing from his shoulders. Another well-known Arabian dragon is the Jinn-dragon, it certainly is said to be a effective creature with the capacity to deliver desires to folks who can seize and manage it.

In Islamic folklore, dragons are frequently associated with the jinn, supernatural creatures manufactured from smokeless fireplace who are believed if you want to granting dreams and appearing magic. According to 3 traditions, dragons had been created with the aid of the jinn to protect the treasures of the earth.

While dragons aren't generally taken into consideration to be a part of Islamic theology, they have got played a huge position within the folklore and mythology of the Arabian Peninsula and surrounding regions for loads of years. Today, they continue to be a famous problem of art work and storytelling in the vicinity.

Bahamut

The fantasy of Bahamut is a famous story from Islamic mythology, and is concept to have originated in historical Arabian folklore. In the story, Bahamut is a big fish or sea monster who is stated to inhabit the depths of the ocean and help the entire universe on its once more.

According to the myth, Bahamut changed into created thru Allah as a effective and majestic creature, with a frame so massive that it would take a bird hundreds of years to fly from one prevent of its body to the alternative. The fish changed into stated to have scales as huge as mountains, and to be so sturdy that no creature, mortal or supernatural, may also need to resist its strength.

In a few variations of the tale, Bahamut is also stated to have a counterpart named Leviathan, each different big sea creature who inhabits the depths of the sea. The creatures are believed to be engaged in an everlasting battle for dominance over the seas and the universe.

Chapter 10: New Zealand

The Taniwha is a mythological creature that originates from the Māori people of New Zealand. It is a creature that is stated to stay in rivers, lakes, and one of a kind our our bodies of water. Taniwha are often depicted as fearsome creatures that can be dangerous to human beings who disturb their territory.

According to Māori mythology, the Taniwha is a father or mother of the waterways, and is stated to have the capability to manipulate the glide of water. They are believed to have a near connection with the environment, and are regularly related to natural disasters at the side of floods and droughts.

There are many precise recollections approximately taniwha, and every iwi (tribe) has their very non-public version of the legend. One of the most famous stories is ready a Taniwha named Tuhirangi, who lived within the Whanganui River. According to the legend, Tuhirangi grow to be a nice Taniwha who helped the network iwi via way of warning them of drawing near floods.

Another well-known tale is set a Taniwha named Ngake, who lived in the Wellington Harbour. Ngake turn out to be said to be a powerful Taniwha who brought about the ocean to surge and created the natural harbour of Wellington.

In modern times, the Taniwha has turn out to be a photograph of Māori cultural identity and is frequently carried out in present day-day paintings and format. The Taniwha is likewise a well-known motif in New Zealand tourism, as many website online site visitors are interested in analyzing extra approximately Māori subculture and mythology.

One of the recollections of the Taniwha is the legend of Taniwha of Porirua, that is a tale from the Wellington place of New Zealand.

According to the legend, a Taniwha named Whataitai lived in the Porirua Harbour. Whataitai modified into said to be a peaceful Taniwha who covered the close by iwi (tribe) and their fishing grounds.

One day, a hard and fast of settlers arrived in the region and started to gather a wharf in the harbour. The creation of the wharf disrupted the flow of the water and disturbed Whataitai's habitat. The Taniwha have end up indignant and commenced to assault the wharf, causing damage and scaring the human beings.

The settlers did now not recognize what to do, so that they went to the community iwi for help. The iwi leaders met with Whataitai and provided a solution. They proposed that the Taniwha take delivery of a ultra-present day domestic in addition up the coast, wherein the water became clearer and the surroundings grow to be better appropriate to Whataitai's needs.

Whataitai agreed to the idea and feature emerge as transported to his new domestic. From that day on, the Taniwha modified into said to have blanketed the iwi and their fishing grounds from afar, and the settlers have been capable to finish the development in their wharf with none in addition interference from Whataitai.

The legend of the Taniwha of Porirua teaches the importance of respecting the surroundings and the creatures that live internal it, similarly to the importance of running together to find out solutions to conflicts.

Aztec

The Aztecs, a pre-Columbian civilization that existed in applicable Mexico from the 14th to the sixteenth century, had a complicated mythology that blanketed quite a few legendary creatures. One such creature come to be the Aztec dragon, which done a fantastic feature in Aztec mythology and life-style.

The Aztec dragon have end up referred to as "Cipactli" or "Cipactonal," because of this "crocodile" or "earth monster." It turned into defined as a creature with an prolonged, serpentine frame, multiple heads, and scales shielding its entire body. According to Aztec mythology, the Aztec dragon end up liable for growing the earth and the sky, and emerge as considered a powerful and fearsome deity.

One of the maximum well-known myths associated with the Aztec dragon is the story of ways it have become killed with the aid of way of the god Huitzilopochtli. According to the parable, the Aztec dragon had wolfed all the previous creations of the gods, and turned into threatening to damage the modern-day international. Huitzilopochtli, the god of struggle and the sun, became selected to slay the dragon and keep the sector. He converted himself into a hummingbird and flew into the dragon's mouth, wherein he then used his weapon, a spear product of lightning, to kill the dragon from the inner.

Another myth related to the Aztec dragon is the story of the goddess Coatlicue, who became the mother of Huitzilopochtli. According to the parable, Coatlicue have grow to be pregnant after she swept up a bundle deal of feathers in a temple. When her kids, on the aspect of Huitzilopochtli, learned of her being pregnant, they have come to be enraged and tried to kill her. However, Huitzilopochtli defended his mother and killed his siblings, which encompass the Aztec dragon, who became one in each of her sons.

The Aztec dragon furthermore played a feature in Aztec artwork and symbolism. It have become regularly depicted in stone carvings, pottery, and outstanding artwork, and modified into associated with thoughts together with creation, destruction, and rebirth. The Aztecs believed that the dragon represented the cyclical nature of the universe, and that its loss of life and rebirth symbolized the ordinary renewal of existence.

While the Aztec dragon is most typically known as Cipactli, there are exclusive dragons and dragon-like creatures in Aztec mythology which might be moreover well simply well worth noting. Here are a few:

Xiuhcoatl - This dragon became called the "fireplace serpent" and modified into associated with the god of fireplace and volcanoes, Xiuhtecuhtli. It become said to have the functionality to shoot flames from its mouth.

Quetzalcoatl - While Quetzalcoatl is not usually portrayed as a dragon, he is often depicted with dragon-like features, together with wings and scales. He become one of the maximum crucial gods in Aztec mythology and changed into associated with know-how, statistics, and creation.

Tlaltecuhtli - This creature is on occasion known as an Aztec dragon because of its serpent-like look. It become associated with the earth and feature end up regularly depicted with its mouth open, organized to eat sacrificial offerings.

Mixcoatl - This god become sometimes depicted as a serpent or dragon and feature become related to the search and struggle.

It's actually well worth noting that the Aztecs did no longer have a selected phrase for "dragon," and the creatures stated above are regularly described the usage of precise phrases together with "serpent," "monster," or "mythical creature."

Xiuhcoatl

Xiuhcoatl is a dragon-like creature in Aztec mythology that become related to hearth, volcanoes, and the sun. According to legend, Xiuhcoatl come to be created through the god of fireplace, Xiuhtecuhtli, as a weapon to wreck the opposite gods.

The myth of Xiuhcoatl tells of a time while the gods were preventing over manage of the universe. Xiuhtecuhtli, who turn out to be irritated at the alternative gods for now not recognizing his strength, created Xiuhcoatl to help him defeat them. The dragon changed into made from a serpent included in obsidian shards, which made it a effective weapon.

When the opposite gods noticed Xiuhcoatl, they have been frightened of its energy and agreed to surrender to Xiuhtecuhtli. However, the god of the sun, Huitzilopochtli, challenged Xiuhcoatl to a war. Huitzilopochtli, who become diagnosed for his bravery and cunning, controlled to trick Xiuhcoatl through throwing a ball of feathers at it. The dragon, mistaking the feathers for prey, lunged on the ball and impaled itself on Huitzilopochtli's spear.

After the conflict, Huitzilopochtli claimed Xiuhcoatl's strength as his very personal and feature grow to be the brand new god of the solar. The Aztecs believed that the dragon's lack of existence and rebirth symbolized the cyclical nature of the universe and the normal renewal of lifestyles.

Xiuhcoatl became a powerful photograph in Aztec way of life and changed into regularly depicted in artwork and shape, especially in association with fire and the sun. Its significance in Aztec mythology presentations the centrality of fireplace and the sun in Aztec cosmology and the cultural importance of war and conquest.

Quetzalcoatl

The fantasy of Quetzalcoatl is one of the most essential and enduring testimonies in Aztec mythology. Quetzalcoatl, whose call way "feathered serpent," turn out to be one of the most essential gods in the Aztec pantheon and turned into related to facts, expertise, advent, and the planet Venus.

According to the myth, Quetzalcoatl come to be born to a virgin named Chimalma, who have become impregnated thru a ball of feathers. As a baby, Quetzalcoatl modified into regarded for his intelligence and piety, and he short have grow to be a respected determine in Aztec society.

As an character, Quetzalcoatl have become a high-quality leader and instructor, referred to for his records and compassion. He taught the Aztecs approximately agriculture, art work, and track, and he emerge as stated to have added peace and prosperity to the land.

However, Quetzalcoatl's peaceful reign become threatened thru the appearance of a rival god, Tezcatlipoca. Tezcatlipoca become jealous of Quetzalcoatl's reputation and sought to undermine his authority thru the use of tricking him into breaking his personal ethical code. Tezcatlipoca glad Quetzalcoatl to bask in pleasures collectively with eating and promiscuity, which went towards Quetzalcoatl's very own teachings.

When Quetzalcoatl located out what had happened, he changed into full of disgrace and disgust and fled into the wilderness. As he wandered, he left inside the lower back of a path of his most prized possessions and treasures, which he gifted to the humans he met along the way. He in the long run arrived on the coast, wherein he built a incredible pyre and set himself on fireside. As he died, he converted into the planet Venus, which stays a outstanding symbol in Aztec subculture to in the mean time.

The delusion of Quetzalcoatl represents the warfare among notable and evil, and the significance of understanding and morality in Aztec society. It is also a testomony to the energy of self-sacrifice and transformation, as Quetzalcoatl's lack of lifestyles and rebirth constitute the cyclical nature of existence and the eternal renewal of the universe.

Tlaltecuhtli

Tlaltecuhtli, moreover known as the "Earth Lord," have become a powerful and fearsome creature in Aztec mythology. It grow to be regularly depicted as a dragon or serpent-like creature, and it turn out to be associated with the earth, fertility, and sacrifice.

According to the myth, Tlaltecuhtli became created with the useful resource of the gods within the course of the creation of the world. When the gods decided to create the earth, they knew they needed a powerful creature to feature its foundation. Tlaltecuhtli come to be made from the body of a goddess, who had sacrificed herself for the great of the area. Her body modified into buried deep in the earth, and from her frame, Tlaltecuhtli emerged.

As the Earth Lord, Tlaltecuhtli had first-rate power over the land and the creatures that inhabited it. It turn out to be said to be so huge that it could swallow whole mountains and cities, and its jaws have been continuously open, prepared to collect services and sacrifices from the Aztec people.

The Aztecs believed that Tlaltecuhtli had to be appeased with offerings and sacrifices that lets in you to ensure the fertility of the land and the prosperity in their civilization. They believed that Tlaltecuhtli's insatiable starvation may also want to simplest be sated thru the blood and hearts of sacrificial sufferers, that have been thrown into its open jaws as offerings.

The fable of Tlaltecuhtli represents the significance of sacrifice and the energy of the earth in Aztec way of life. It moreover reflects the Aztecs' understanding of the cyclical nature of life and the interconnectedness of all things, as

Tlaltecuhtli was seen as every the muse of the earth and the recipient of the sacrifices that ensured its endured fertility and prosperity.

Mixcoatl

Mixcoatl have grow to be a god of looking, battle, and the Milky Way in Aztec mythology. According to the parable, he became the son of a goddess named Coatlicue, and he changed into born as a totally grown person.

Mixcoatl became diagnosed for his skills in looking, and he have become often depicted wearing a bow and arrow. He have become moreover related to the night time time sky and the Milky Way, which the Aztecs believed to be a searching path that precipitated the afterlife.

In one model of the myth, Mixcoatl is stated to have led the Aztecs on a great migration from their vicinity of beginning location of Aztlan to their new home in the Valley of

Mexico. Along the way, he taught the Aztecs about looking, agriculture, and struggle, and he helped them conquer many obstacles and disturbing situations.

In each one of a kind version of the myth, Mixcoatl is stated to were the leader of a hard and fast of warriors who had been tasked with searching down and taking pictures the god Tezcatlipoca. With his looking abilities and the assist of his allies, Mixcoatl turned into able to seize Tezcatlipoca and convey him to justice.

The fantasy of Mixcoatl represents the importance of searching and struggle in Aztec manner of lifestyles, as well as the concept of migration and the search for a present day domestic. It moreover presentations the Aztecs' reverence for the herbal global and the night time time sky, and their belief inside the interconnectedness of all topics.

Chapter 11: Australia

There is no specific fable of an Australian dragon. However, there are numerous mythological creatures from Australian Aboriginal cultures that might be interpreted as dragon-like beings.

One such creature is the Rainbow Serpent that may be a powerful and sacred being this is said to have created the panorama of Australia. The Rainbow Serpent is frequently depicted as an prolonged, serpentine creature with colorful scales and the capability to govern water and the climate.

Another creature is the Bunyip, it's a big, fearsome monster said to inhabit swamps, billabongs, and different our our bodies of water. The Bunyip is frequently described as having the frame of an ox, the pinnacle of a horse, and the tail of a platypus, even though its look varies relying on the place and cultural agency.

While those creatures won't be particularly known as dragons, they share many developments with traditional dragon myths, which encompass their powerful and often fearsome nature, their connection to water and the herbal worldwide, and their critical role in Australian folklore and mythology.

The Rainbow Serpent

The Rainbow Serpent is a distinguished mythological parent in lots of Aboriginal cultures in Australia. It is a effective and sacred being this is believed to have created the panorama, animals, and people of Australia.

According to the myth, the Rainbow Serpent emerged from the Dreamtime, a length of introduction that befell in the historic beyond. The Serpent traveled at some level in the land, carving out valleys and rivers as it went. It furthermore created animals and plants, and it's miles stated that the

Serpent's moves created the winding paths of the rivers and the rainbows that arch in the course of the sky.

In many Aboriginal cultures, the Rainbow Serpent is seen as a photograph of fertility, advent, and renewal. Its powers include manage over the factors of water and climate, and it's far regularly associated with rain, thunder, and lightning. The Serpent is likewise a figure of religious significance, and its presence is believed to preserve recuperation and blessings.

The tale of the Rainbow Serpent has been handed down thru generations of Aboriginal human beings through oral way of life and paintings. Many Aboriginal artistic endeavors characteristic depictions of the Serpent, often in the shape of intricate, colourful designs that represent the Serpent's scales or the varieties of the land that it created.

Today, the Rainbow Serpent stays an important image of Aboriginal manner of lifestyles and spirituality, and its tale remains a powerful reminder of the connection amongst people, the herbal global, and the forces of creation.

The Bunyip

The Bunyip is a mythological creature from Australian Aboriginal folklore. It is frequently defined as a big, fearsome monster that inhabits swamps, billabongs, and other our bodies of water for the duration of Australia.

The appearance of the Bunyip varies relying on the area and cultural agency, however it also includes depicted as having a mixture of numerous animal talents. Some descriptions propose that it has the frame of an ox, the head of a horse, and the tail of a platypus, while others describe it as a big, furry creature with a massive, toothy mouth.

The Bunyip is understood for its loud, eerie cry, that is stated to be heard at night time time close to our our bodies of water. It is thought to be a nocturnal creature that feeds on fish, water birds, and different animals that live in or close to the water.

The legend of the Bunyip has been a part of Aboriginal manner of lifestyles for masses of years, and it stays a well-known scenario of folklore and storytelling in Australia these days. Some humans believe that the Bunyip is a real creature that has however to be positioned through technological expertise, at the identical time as others view it as a actually legendary creature and now not using a foundation in fact.

Regardless of its life, the Bunyip remains an critical a part of Australian folklore, and its legend serves as a reminder of the united states of a's rich cultural heritage and the long-lasting power of delusion and storytelling.

Bhutan

In Bhutanese mythology, dragons are known as "Druk" and are taken into consideration to be the embodiment of the u . S .'s spirit. The dragon is one of the maximum important symbols of Bhutan, and it's miles found at the u . S . A .'s countrywide flag, forex, and legitimate seals.

According to legend, the Druk have become selected because the countrywide photograph of Bhutan via the usage of a Tibetan lama named Drukpa Kunley inside the 15th century. Drukpa Kunley is also known as the "Divine Madman" and is said to have tamed a demon that were terrorizing the community populace. He is stated to have shot an arrow into the air, and the demon emerge as transformed right into a rock that is however visible these days inside the Punakha valley.

In Bhutanese mythology, the Druk is a powerful creature this is said to stay inside the Himalayan mountains. It is frequently depicted with 4 limbs, a long tail, and wings, and it is said to breathe fireplace. The Druk is taken into consideration to be a protector of the Bhutanese human beings and is said to have the electricity to carry rain and control the weather.

The Druk is also related to the Bhutanese royal circle of relatives. The cutting-edge king of Bhutan, Jigme Khesar Namgyel Wangchuck, is referred to as the "Dragon King" and is seen as a reincarnation of the dragon. The dragon is also a common problem remember in Bhutanese paintings and structure, and lots of buildings in the u . S . A . Characteristic dragon motifs.

Overall, the Druk is an crucial photo of Bhutanese way of existence and identity, and it reflects the u . S .'s unique mythology and spiritual traditions.

The Druk

The Druk, additionally referred to as the "Thunder Dragon," is a extremely good image in Bhutanese lifestyle and is taken into consideration to be the embodiment of the us of the us's identification. The dragon is often depicted in Bhutanese art and architecture, and it's far featured on the united states of the usa's countrywide flag, foreign exchange, and actual seals.

The Druk is commonly depicted with 4 limbs, an extended tail, and wings, and it is said to breathe fireside. It is considered to be a effective creature with the capability to manipulate the weather, deliver rain, and protect the Bhutanese people. The Druk is also related to the Bhutanese royal circle of relatives, and the present day king of Bhutan, Jigme Khesar Namgyel Wangchuck, is called the "Dragon King."

According to legend, the Druk emerge as on the begin a demon that terrorized the people of Bhutan. It become tamed with the useful resource of a Buddhist saint named Guru Rinpoche, who is stated to have flown at the lower lower back of a tiger to subdue the demon. The Druk modified into then converted right right right into a protector of the Bhutanese human beings and end up adopted because of the fact the national photograph of Bhutan.

The Druk is likewise associated with the Buddhist concept of "Drukpa," which refers back to the "lineage of the dragon." This lineage is primarily based mostly on the teachings of the Drukpa Kagyu university of Buddhism, which grow to be founded within the 12th century with the aid of Tsangpa Gyare, a Buddhist grasp who is stated to have had a vision of nine dragons.

Overall, the Druk is an vital part of Bhutanese way of life and mythology, and it displays america of a's particular religious traditions and identity. It is a photograph of energy, electricity, and protection, and it's far respected via the Bhutanese human beings as a sacred and auspicious creature.

Yemen

The dragon delusion of Yemen is targeted round a creature called the Qahārāt. According to Yemeni folklore, the Qahārāt is a fearsome dragon that lived within the mountains of Yemen, preying on vacationers and villagers who lived close by.

The Qahārāt became said to be pretty big, with scales as tough as iron and sharp claws and teeth. It become moreover believed to have the energy of flight, making it tough to interrupt out as quickly because it had targeted its prey.

In some versions of the myth, the Qahārāt changed into believed to be invulnerable to any form of attack. However, in others, it changed into said that the first-rate manner to defeat the dragon was to pierce its coronary coronary heart with a totally particular sword, known as the "Zulfikar".

Despite its fearsome popularity, the Qahārāt have emerge as additionally believed to non-public extremely good knowledge and information. Some reminiscences suggest that it have become able to speaking human languages or maybe had the potential to are looking forward to the destiny.

The legend of the Qahārāt has been handed down through generations in Yemen, and remains a popular challenge of folklore and storytelling in the u . S ..

One of the most famous recollections of the Qahārāt tells of a courageous warrior named Amr who got right down to defeat the fearsome dragon and shop his village from its reign of terror.

Amr changed proper right into a professional swordsman and had professional for years to combat the Qahārāt. He set out on his adventure together with his depended on horse and sword, decided to face the dragon.

After severa days of adventure, Amr ultimately reached the mountains where the Qahārāt changed into said to stay. He climbed the steep cliffs and reached the dragon's lair, in which he decided the creature sleeping.

Amr approached the dragon carefully, but his sword didn't pierce the Qahārāt's scales. The dragon wakened and attacked Amr, but the warrior modified into brief and agile, dodging the dragon's attacks and setting it with his sword.

The war amongst Amr and the Qahārāt raged on for hours, with every aspects causing damage on the opportunity. Finally, Amr controlled to land a deadly blow, piercing the dragon's coronary coronary heart along together with his sword.

With the Qahārāt defeated, Amr again to his village a hero. He have end up celebrated for his bravery and hailed as a champion who had saved his humans from the fearsome dragon.

The story of Amr and the Qahārāt continues to be suggested in Yemen, inspiring courage and bravery in the face of chance.

Ethiopia

In Ethiopian mythology, there may be a powerful dragon referred to as the "Bida". The Bida is said to be a massive serpent that is capable of inflicting droughts and destroying plant life. It is likewise believed to have the energy to govern the rain and is associated with thunder and lightning.

According to Ethiopian legend, the Bida lived in the Lake Tana place of Ethiopia and became feared via the local people. It modified into said to be so huge that it is able to swallow entire villages complete, and its breath have become stated to be so warm that it may scorch the earth.

In order to soothe the Bida and prevent it from inflicting destruction, the area people would possibly perform complex ceremonies and provide sacrifices to the creature. These services had been regularly in the form of food, farm animals, and precise valuable gadgets.

The Bida is likewise believed to have the electricity to shape-shift and take at the form of a lovely lady, luring guys to their deaths. In a few versions of the myth, a brave hero is capable of defeat the Bida with the useful resource of outwitting it or via the usage of a powerful magical weapon.

Overall, the Bida is an vital part of Ethiopian mythology and represents the electricity of nature and the need to admire and appease the forces that manipulate it. The tale of the Bida furthermore demonstrates the significance of courage and wit inside the face of first-rate chance.

Chapter 12: The Bida

There are several variations of the tale of the Bida in Ethiopian mythology. Here is one common model:

Long inside the past, inside the Lake Tana place of Ethiopia, there lived a effective dragon named the Bida. The Bida come to be a huge serpent that became feared by using the local community, because it became able to causing droughts, destroying flowers, and swallowing entire villages complete. Its breath have come to be said to be so heat that it can scorch the earth.

The community humans believed that the Bida can also want to govern the rain and lightning, and they carried out tricky ceremonies and furnished sacrifices to appease the creature. These offerings had been often in the form of meals, livestock, and extraordinary precious gadgets.

One day, a courageous hero named Qemal were given all the way all the way down to defeat the Bida and maintain his people from its destructive power. He armed himself with a effective magical weapon, a sword cast from the best metals and blessed with the useful resource of the gods.

Qemal journeyed to the beaches of Lake Tana, in which he determined the Bida resting in the water. He known as out to the creature, tough it to a struggle. The Bida fashionable the challenge and emerged from the water, its big frame coiling spherical Qemal.

The battle among Qemal and the Bida come to be lengthy and fierce. Qemal swung his sword with all his would possibly, however the Bida end up too effective for him to defeat. Just while it regarded that every one preference changed into misplaced, Qemal remembered a chunk of recommendation he had acquired from an antique smart guy.

The practical guy had knowledgeable him that the Bida may want to best be defeated if it had been struck inside the correct middle of its forehead. Qemal aimed his sword cautiously and struck the Bida within the forehead with all his can also. The dragon let loose a exceptional roar and fell to the floor, defeated.

Qemal returned to his village, hailed as a hero for defeating the fearsome Bida. The local people celebrated and thanked Qemal for his bravery, and the dragon changed into in no manner seen all another time. The tale of Qemal and the Bida is still informed in Ethiopia as a story of braveness, electricity, and the triumph of proper over evil.

Bulgaria

In Bulgarian mythology, the dragon is called "zmei" or "lamia". The zmei is depicted as a winged serpent with a lion's head and fiery eyes. It is stated to live in deep underground caves or inside the mountains and is notion to personal magical powers.

According to Bulgarian legends, the zmei is often related to the underworld and represents chaos and destruction. It is also believed to be a photo of fertility and rebirth. In some memories, the zmei is depicted as a protector of princesses and is often depicted as a shapeshifter who can tackle human shape.

One well-known Bulgarian fable tells the tale of a hero named Krali Marko, who want to defeat a effective zmei that has been terrorizing a close-by village. Krali Marko tricks the zmei via convincing it to swallow a crimson-warm iron rod, which reasons the dragon to blow up into 1000 pieces.

Another myth includes a zmei that kidnaps a princess and takes her to its lair in the mountains. The hero, Samodiva, need to rescue the princess and defeat the zmei thru using his wit and cunning.

Overall, the dragon myths of Bulgaria are charming and rich in symbolism, offering notion into the ancient beliefs and way of existence of the Bulgarian human beings.

In Bulgarian mythology, the zmei (змей in Cyrillic) is a dragon-like creature with serpent talents, wings, and a lion's head. It is assumed to be a effective and magical creature, able to bringing each appropriate and terrible fortune to human beings.

The zmei is frequently depicted as a parent of hidden treasure or magical artifacts, along with a ring or a sword. It is likewise associated with natural phenomena which encompass thunderstorms and earthquakes.

In some Bulgarian myths, the zmei is a shapeshifter and can take at the shape of a human. It is frequently portrayed as a trickster, using its cunning to mislead human beings and now and again playing pranks on them.

There also are reminiscences of zmei who kidnap younger maidens, often princesses, and take them to their lair inside the mountains. In the ones stories, a brave hero want to rescue the princess and defeat the zmei, normally by outwitting it or using a mystical weapon.

The zmei is deeply rooted in Bulgarian folklore and stays an essential cultural symbol in recent times. Its image may be seen in hundreds of traditional Bulgarian crafts and artwork, and the creature is frequently featured in oldsters songs and dances.

Overall, the zmei is a fascinating and complicated creature in Bulgarian mythology, representing each the electricity of nature and the cunning of the human spirit.

One of the most famous memories concerning a zmei in Bulgarian mythology is the story of Krali Marko and the Zmei. Krali

Marko is a heroic determine in Bulgarian folklore, regarded for his power and foxy, and the zmei is one among his most ambitious adversaries.

The story is going that a zmei become terrorizing a close-by village, stealing the villagers' cattle and causing destruction anywhere it went. The villagers appealed to Krali Marko for assist, and he got down to defeat the zmei.

When Krali Marko located the zmei, the creature challenged him to a sequence of contests. First, the zmei challenged Krali Marko to a wrestling in shape, but Krali Marko emerged successful. Then, the zmei challenged him to a exercise of energy, but Krali Marko over again emerged successful.

Finally, the zmei challenged Krali Marko to a undertaking of wits. The zmei said he may swallow a crimson-hot iron rod and then ask Krali Marko a chain of riddles. If Krali Marko couldn't answer the riddles, the zmei would kill him. But if Krali Marko replied effectively, the zmei may also launch the villagers' cattle and in no way go back to the village again.

Krali Marko agreed to the task, and the zmei swallowed the purple-warmth iron rod. He then requested Krali Marko a sequence of hard riddles, however Krali Marko answered all of them correctly. In the prevent, the zmei have end up forced to launch the villagers' livestock and depart the village by myself.

This tale is often cautioned in Bulgaria as a story of braveness, energy, and wit winning over evil. It additionally demonstrates the electricity and foxy of the zmei, that is considered one of the most bold creatures in Bulgarian folklore.

Myanmar

In Myanmar (previously called Burma), the dragon is called the "naga" and performs a large function within the u.S. Of america's mythology and way of life. The naga is considered a effective and mythical creature this is believed to have existed for plenty of years.

According to the parable, the naga is a serpent-like creature with a dragon's head and a frame this is frequently depicted with scales and a serpentine tail. The naga is related to water and is idea to live in rivers, lakes, and seas. It is regularly depicted in art work and sculptures in Myanmar, and is visible as a image of energy, strength, and protection.

In Myanmar, the naga is frequently related to Buddhism and is belief to have performed a position inside the existence of the Buddha. According to legend, at the same time as the Buddha modified into meditating underneath a tree, a naga king protected him from a typhoon through way of using protecting the Buddha along collectively with his hood.

The naga is also believed to have achieved a position in the founding of Myanmar. According to the myth, the primary king of Myanmar have end up a prince who became given a mystical sword through manner of the naga king. The sword turned into stated to were used to defeat the prince's enemies and installation the number one nation of Myanmar.

Today, the naga continues to play an vital function in Myanmar's life-style and is regularly depicted in artwork, literature, and religious ceremonies. The naga is also considered a photo of countrywide identification and is featured at the u . S . A .'s flag.

The naga is a legendary creature this is high-quality in the cultures and religions of numerous Southeast Asian global locations, such as Myanmar, Thailand, Laos, Cambodia, and Indonesia. In addition to its affiliation with Buddhism in Myanmar, the naga is also frequently related to Hinduism and is believed to be a powerful and supernatural being.

In many cultures, the naga is related to water and is often depicted as a serpent or dragon-like creature with a hood or more than one heads. The naga is idea to be a protector of the natural international, and is often associated with rivers, lakes, and different our our bodies of water. In some legends, the naga is also believed to be a dad or mum of treasures and is associated with wealth and prosperity.

In addition to its association with Buddhism and Hinduism, the naga is likewise an important discern in animist notion systems. Animism is a perception machine that is based totally totally on the concept that the entirety within the global, collectively with plant life, animals, and herbal devices, has a spirit or soul. In animist ideals, the naga is regularly visible as a powerful and benevolent spirit that may offer safety and steerage.

The naga is frequently depicted in art work and sculpture within the course of Southeast Asia, and is a commonplace situation rely in religious iconography. In a few cultures, the naga is also believed to be a form-shifter, capable of remodel proper right into a human or animal shape. Overall, the naga is a charming and multifaceted creature that performs an essential feature in the mythology and way of lifestyles of many Southeast Asian global locations.

Azerbaijan

One of the most famous dragon myths of Azerbaijan is the tale of a creature called Zilant.

According to the legend, Zilant turn out to be a dragon with the body of a snake and wings of a hen. It come to be stated to stay in a cave underneath the city of Baku and terrorized the human beings dwelling inside the surrounding regions.

One day, a brave warrior named Rustam became summoned to defeat the dragon. He set out on a journey and after several days of visiting, he arrived at the doorway of Zilant's cave.

Rustam drew his sword and entered the cave, wherein he decided Zilant dozing. He approached the dragon quietly and with one fast stroke, he decapitated the beast.

The humans of Azerbaijan celebrated Rustam's bravery and erected a statue in his honor in the town of Baku. The dragon's head was moreover displayed as a trophy.

Today, Zilant is a image of Azerbaijan and is featured on the u . S .'s coat of palms. The legend of Zilant additionally serves as a reminder of the braveness and bravery of the Azerbaijani people.

Chapter 13: Latvia

There are numerous dragon myths in Latvian folklore, but perhaps the most well-known is the tale of "Lāčplēsis," which interprets to "Bear-slayer" in English.

According to the myth, Lāčplēsis grow to be a effective warrior who have become raised with the resource of a bear inside the forests of Latvia. As he grew older, he became identified for his wonderful power and bravado, and turned into in the end known as upon with the useful resource of the ruler of Latvia to help defend the u . S . A . From invading German knights.

During the battle, Lāčplēsis encountered a fierce dragon which have been terrorizing the Latvian people. The dragon had seven heads and feature grow to be stated to breathe hearth and poison. Lāčplēsis fought bravely toward the dragon, in the long run dealing with to reduce off all seven of its heads. However, as he attempted to supply the final blow, the dragon's tail lashed out and struck him, inflicting him to fall unconscious.

When Lāčplēsis awakened, he positioned that the dragon's blood had given him superhuman energy and talents. He used this newfound power to defeat the German knights and feature turn out to be a country wide hero in Latvia.

Today, Lāčplēsis is widely known as a photo of Latvian strength and resilience, and the dragon is frequently used as a metaphor for the demanding situations that the Latvian people have overcome throughout their facts.

Here are some extra Latvian dragon myths:

"Pūķis" - This is a dragon-like creature in Latvian folklore this is stated to stay in the water. It has an extended, serpentine frame and is regularly depicted with more than one heads. Pūķis is concept for its capacity to manipulate the weather, in particular storms and floods.

"Sarkanā pūķe" - This is another water dragon in Latvian mythology, however it's also depicted as a red serpent with wings. According to legend, the Sarkanā pūķe could probably every so often emerge from the water and attack boats passing with the aid of using the use of.

"Zaķu pūķis" - This is a dragon this is said to be the offspring of a hare and a serpent. It is normally depicted as a small, winged creature with rabbit-like abilities. According to fable, the Zaķu pūķis is a mischievous creature that enjoys gambling tips on humans.

"Lauvas sārts" - This is a dragon this is stated to have the body of a lion and the wings of a dragon. It is usually depicted as a fiery red colour and is known for its fierce mood. According to legend, the Lauvas sārts changed into one of the guardians of the Latvian underworld, and will from time to time emerge to wreak havoc at the residing international.

Pūķis

Pūķis is a dragon-like creature in Latvian folklore this is stated to live within the water. It is often depicted as an prolonged, serpentine creature with more than one heads, and is understood for its potential to control the weather, specifically storms and floods.

According to legend, Pūķis changed proper right into a effective deity who managed the waters and the climate. In some versions of the parable, Pūķis emerge as a benevolent deity who used his powers to supply rain to the flowers and hold the rivers flowing. However, in remarkable versions of the myth, Pūķis end up a malevolent creature who could cause storms and floods to punish people who angered him.

In Latvian folklore, Pūķis have turn out to be frequently associated with the spring and summer season months, whilst the climate changed into most unpredictable. It come to be believed that Pūķis have to emerge from the water in the course of thunderstorms, and that his roars may be heard during the geographical region.

Today, Pūķis remains an crucial photograph in Latvian lifestyle, and is frequently used to symbolize the energy and unpredictability of nature. Many Latvians nevertheless agree with inside the delusion of Pūķis, and a few even make offerings to him within the hopes of appeasing his wrath and ensuring appropriate weather for his or her plants.

Sarkanā pūķe

Sarkanā pūķe, moreover referred to as the Red Dragoness, is a dragon-like creature in Latvian mythology this is stated to stay in the water. It is commonly depicted as a pink serpent with wings, and is concept for its fierce mood and capability to cause chaos and destruction.

According to legend, Sarkanā pūķe may want to from time to time emerge from the water and attack boats passing through the use of. It come to be believed that the dragoness could goal boats wearing those who had angered her, and might sink them at the side of her powerful tail. In some variations of the parable, Sarkanā pūķe end up said to be the offspring of a dragon and a serpent, giving her the capacity to breathe fireside and manipulate the climate.

Despite her fearsome reputation, Sarkanā pūķe have emerge as moreover visible as a image of fertility and rebirth. In Latvian folklore, it grow to be believed that the dragoness might upward thrust from the water sooner or later of the springtime, bringing new existence to the land and making sure a bountiful harvest.

Today, Sarkanā pūķe remains a famous figure in Latvian way of life, and is often depicted in art work and literature. She is seen as a powerful and unpredictable strain of nature, able to both destruction and creation.

Zaķu pūķis

Zaķu pūķis is a dragon-like creature in Latvian folklore that is stated to be the offspring of a hare and a serpent. It is generally depicted as a small, winged creature with rabbit-like features, and is thought for its mischievous nature.

According to legend, Zaķu pūķis may additionally want to sometimes emerge from its hiding region inside the wooded region and play hints on people. It have emerge as said to have the energy to shape-shift into precise paperwork, and may use this functionality to confuse and confound its victims. In some variations of the parable, Zaķu pūķis have become said to be the mum or dad of hidden treasure, and may trap human beings into the woodland as a manner to scouse borrow their valuables.

Despite its recognition as a trickster, Zaķu pūķis became furthermore visible as a protector of the woodland and its populace. In some versions of the myth, the creature might use its powers to guard the wooded location from folks that would in all likelihood are seeking out to damage it.

Today, Zaķu pūķis is still a famous determine in Latvian manner of lifestyles, and is frequently depicted in artwork and literature. It is visible as a photo of the playful and mischievous aspect of nature, and is a reminder to treat the environment with recognize and care.

Lauvas sārts

Lauvas sārts is a dragon-like creature in Latvian mythology this is stated to have the frame of a lion and the wings of a dragon. It is generally depicted as a fiery red color and is belief for its fierce mood and power.

According to legend, Lauvas sārts modified into one of the guardians of the Latvian underworld, and will from time to time emerge to wreak havoc at the residing international. It have come to be believed that the creature may also breathe fireside and motive earthquakes, and that its roar can be heard during the geographical region.

Despite its fearsome popularity, Lauvas sārts have become moreover visible as a image of strength and braveness. In a few versions of the parable, the creature might come to the resource of these in want, the use of its powerful claws and teeth to guard them from threat.

Today, Lauvas sārts continues to be a well-known determine in Latvian lifestyle, and is frequently depicted in art and literature. It is visible as a reminder of the strength of nature, and the need to comprehend and honor the forces that govern the arena round us.